Applications and Metrology at Nanometer Scale 1

Reliability of Multiphysical Systems Set
coordinated by
Abdelkhalak El Hami

Volume 9

Applications and Metrology at Nanometer Scale 1

Smart Materials, Electromagnetic Waves and Uncertainties

Pierre-Richard Dahoo
Philippe Pougnet
Abdelkhalak El Hami

WILEY

First published 2021 in Great Britain and the United States by ISTE Ltd and John Wiley & Sons, Inc.

Apart from any fair dealing for the purposes of research or private study, or criticism or review, as permitted under the Copyright, Designs and Patents Act 1988, this publication may only be reproduced, stored or transmitted, in any form or by any means, with the prior permission in writing of the publishers, or in the case of reprographic reproduction in accordance with the terms and licenses issued by the CLA. Enquiries concerning reproduction outside these terms should be sent to the publishers at the undermentioned address:

ISTE Ltd
27-37 St George's Road
London SW19 4EU
UK

www.iste.co.uk

John Wiley & Sons, Inc.
111 River Street
Hoboken, NJ 07030
USA

www.wiley.com

© ISTE Ltd 2021
The rights of Pierre-Richard Dahoo, Philippe Pougnet and Abdelkhalak El Hami to be identified as the authors of this work have been asserted by them in accordance with the Copyright, Designs and Patents Act 1988.

Library of Congress Control Number: 2020946679

British Library Cataloguing-in-Publication Data
A CIP record for this book is available from the British Library
ISBN 978-1-78630-640-1

Contents

Preface ix

Introduction xiii

Chapter 1. Nanometer Scale 1

    1.1. Introduction 1
    1.2. Sample elaboration 6
        1.2.1. Physical and chemical method: spin coating 9
        1.2.2. Physical method: cathode sputtering 12
        1.2.3. Physical method: laser ablation 14
    1.3. Characterization of samples 20
        1.3.1. Scanning electron microscope 26
        1.3.2. Atomic force microscope 30
        1.3.3. Infrared spectroscopy (FTIR/ATR) 33
    1.4. Conclusion 45
    1.5. Appendix: light ray propagation 46

**Chapter 2. Statistical Tools to Reduce the Effect of Design
Uncertainties** 51

    2.1. Introduction 51
    2.2. Review of fundamental definitions in probability theory 52
        2.2.1. Definitions and properties 52
        2.2.2. Random variables 54
        2.2.3. Random vectors 55
        2.2.4. Static moments 56
        2.2.5. Normal probability functions 60
        2.2.6. Uniform probability function 61

vi   Applications and Metrology at Nanometer Scale 1

2.3. Random process and random field 62
2.4. Mathematical formulation of the model 64
2.5. Reliability-based approach. 65
   2.5.1. Monte Carlo method 65
   2.5.2. Perturbation method 66
   2.5.3. Polynomial chaos method 70
2.6. Design of experiments method 71
   2.6.1. Principle 71
   2.6.2. Taguchi method. 72
2.7. Set-based approach 76
   2.7.1. The interval method 77
   2.7.2. Fuzzy logic-based method. 79
2.8. Analysis in terms of main components. 82
   2.8.1. Description of the approach. 82
   2.8.2. Mathematical basis 83
   2.8.3. Interpretation of results 84
2.9. Applications 84
   2.9.1. Rod mesh. 84
   2.9.2. Example of a linear oscillator. 88
2.10. Conclusion 90

Chapter 3. Electromagnetic Waves and Their Applications.   91

3.1. Introduction. 91
3.2. Characteristics of the energy carried by an electromagnetic wave . . . 94
3.3. The energy of a plane monochromatic electromagnetic wave 98
   3.3.1. Answer to question 1. 99
   3.3.2. Answer to question 2. 100
   3.3.3. Answer to question 3. 100
   3.3.4. Answer to question 4. 101
   3.3.5. Answer to question 5. 101
   3.3.6. Answer to question 6. 103
3.4. Rectangular waveguide as a high-pass frequency filter 103
   3.4.1. Answer to question 1. 105
   3.4.2. Answer to question 2. 107
   3.4.3. Answer to question 3. 108
   3.4.4. Answer to question 4. 108
   3.4.5. Answer to question 5. 109
   3.4.6. Answer to question 6. 110
   3.4.7. Answer to question 7. 111
   3.4.8. Answer to question 8. 111
   3.4.9. Answer to question 9. 111
   3.4.10. Answer to question 10 112

Contents vii

3.4.11. Answer to question 11 112
3.4.12. Answer to question 12 112
3.4.13. Answer to question 13 113
3.4.14. Answer to question 14 113
3.4.15. Answer to question 15 114
3.5. Characteristics of microwave antennas 114
3.5.1. Introduction to antennas 115
3.5.2. Radiation of a wire antenna 122
3.6. Characteristics of networks of microwave antennas 134
3.6.1. Introduction to networks of microwave antennas 134
3.6.2. Radiation of antenna networks 137

Chapter 4. Smart Materials 147

4.1. Introduction 147
4.2. Smart systems and materials 150
4.3. Thermodynamics of couplings in active materials 158
4.3.1. Thermo-mechanical and thermoelastic coupling 158
4.3.2. Multiphysics couplings 172
4.4. Exercises on the application of active materials 184
4.4.1. Strain tensor for 2D thin films 184
4.4.2. A piezoelectric accelerometer 190
4.4.3. Piezoelectric transducer 193
4.4.4. Piezoelectric sensor 198
4.5. Appendix: crystal symmetry 202

Appendix 205

References 211

Index 217

Preface

The various actions taken worldwide in support of sustainable development and greenhouse gas emission control have led to increasingly restrictive regulations. Manufacturers in the automotive field have thus developed innovative mechatronic systems, enabling various vehicle functions to go electric. Confronted with the globalization of exchanges that has generated stronger competition and a surge of new product performances, companies in the sector of embedded mechatronic systems are developing new products at an increasingly faster rate.

In other domains, to achieve volume or mass reduction or curb energy dissipation, manufacturers of mechatronic systems are developing new assembly methods based on multimaterials (e.g. composite materials, hybrids) or on innovative nanomaterials (e.g. carbon nanotubes). Modeling is essential to reduce product development cost, shrink time to market, understand failure mechanisms occurring in the operating conditions and optimize design before launching production. The reliability-based design optimization (RBDO) method is a modeling approach that optimizes design, guarantees the required high level of performance and takes into account the variability in the manufacturing process as well as the climatic variability in the use conditions. The efficiency of RBDO, however, depends on a solid understanding of the failure mechanisms caused by aging.

Creating a model of a dynamic system often involves developing a simplified model of its behavior based on realistic hypotheses and on the key parameters that are required for its functioning. The dynamic behavior of this modeled system is ruled by partial differential equations (PDEs). The model is then improved by introducing elements or parameters that were not initially included and by improving the set of PDEs (nonlinearity, coupling, etc.), in order to obtain a model that closely represents the reality of operating systems and provides pertinent simulation results.

The theoretical models that are based on the fundamental laws of physics use a *bottom-up* approach. These models can be studied using analytical or numerical methods. When experiments can be implemented, simulation results are compared to experimental results. It is also possible to use experimental methods and a *top-down* approach to build a database of the response of the system to applied stresses. These data are then analyzed by comparing them to the response of theoretical or empirical models. In all the cases, there is a degree of uncertainty in the statistical analysis of the data, which leads to predictions with a margin of error. The lower the margin of error, the closer the predictions are to reality, leading to a sound understanding of the functionalities of working materials.

As Book 9 of the "Reliability of Multiphysical Systems Set", this book is designed to provide applications for Book 2 in the set, entitled *Nanometer-scale Defect Detection Using Polarized Light*. This is achieved by describing the experimental and theoretical methods developed in fundamental research laboratories to understand the physics or chemical processes, which at the nanometer scale are at the origin of the remarkable properties of the materials introduced in innovative technological devices. It presents optical techniques based on polarized light, which are used to characterize interface and bulk material defects that have an impact on the performance of nanodevices. It also describes how mechanical properties of nanomaterials can be determined using theoretical models and by the analysis of experimental results and their uncertainties.

This book is intended for students at master and doctoral levels, teaching academics and researchers in materials science, physics engineering and experimental study, as well as R&D and manufacturing engineers of large groups and SMEs in the field of electronics, mechatronics, or optical or electronic materials.

Chapter 1 provides a historical overview of the development of nanosciences and nanotechnologies and describes the challenges encountered when working on the nanometric scale, such as finding new ways to measure the physical properties of nanomaterials. It provides an overview of the techniques used for manufacturing nanomaterials: spin coating, cathode sputtering and laser ablation. Four characterization and failure analysis techniques adapted to nanotechnologies are presented: transmission electron microscopy (TEM), scanning electron microscopy (SEM), atomic force microscopy (AFM) and attenuated total reflection (ATR) infrared spectroscopy.

Chapter 2 describes how to manage system variable uncertainties in the design process. The objective is to obtain a design that meets the performance requirements, has a stable response when design parameters vary randomly and respects a threshold of minimal performance for a given confidence level. Several methods for analyzing the effect on the output of uncertainties in the system input parameters are presented with practical applications: a probabilistic approach, interval analysis, a fuzzy logic method, designs of experiments and principal component analysis.

Chapter 3 is dedicated to various applications of electromagnetic waves. After a quick summary of the main characteristics of electromagnetic waves and microwave antenna theory, the following applications are studied: energy of a monochromatic plane wave, properties of a rectangular waveguide, performance of a wire antenna and antenna networks. These applications facilitate the understanding of the operation of antennas for the fifth generation (5G) of mobile telecommunication systems.

Chapter 4 deals with functional materials employed in intelligent systems. The main characteristic of these intelligent materials is the coupling of their various physical properties. Thermodynamic coupling and multiphysics coupling are studied for piezoelectric, magnetostrictive and shape memory materials. Application exercises are provided for the deformations of a plate-like thin layer, a piezoelectric accelerometer, a piezoelectric transducer and a piezoelectric sensor.

Pierre Richard DAHOO
Philippe POUGNET
Abdelkhalak EL HAMI
October 2020

Introduction

The scientific study of measurement is known as metrology. Any measure is based on a universally accepted standard and any measuring process is prone to uncertainty. In engineering science, measurement concerns various types of parameters. Legal metrology is imposed by a regulatory framework that the manufactured product must respect. Technical or scientific metrology involves the methods used to measure the technical characteristics of the manufactured product. In engineering sciences, measurement concerns various types of parameters. In a more general context of a systemic approach, metrology should also be considered in connection with other indicators of the production system. These measures enable the follow-up and development of the processes implemented for ensuring and optimizing product quality or reducing failure so that it meets client expectations. The ability of a product to meet quality and reliability expectations can be addressed in the design stage, according to a RBDO (Reliability-Based Design Optimization) approach described in the "Reliability of Multiphysical Systems Set Book 2", entitled *Nanometer-scale Defect Detection Using Polarized Light*. More generally, RBDO makes it possible to consider the uncertain parameters of manufacturing processes, measurement and operational conditions in order to optimize the manufacturing process, the design parameters and the overall quality of the product.

Reliability of Multiphysical Systems Set Book 2 focused on three levels of design for manufacturing an industrial product:

– Numerical methods developed in engineering from mathematical models and theories in order to optimize product quality from its design according to RBDO. This methodology is a source of applications in engineering science intended to address optimization problems in the industrial field.

– Experimental methods developed in fundamental research relying on the light–matter interaction and on simulation-based analysis using theoretical models in order to make nanometer-scale measurements and conduct the analysis. These methods are used in nanosciences for the elaboration of knowledge leading to nanotechnologies.

– Finally, the application of these two approaches in the example presented in Chapter 9 of Reliability of Multiphysical Systems Set Book 2 (*Nanometer-scale Defect Detection Using Polarized Light*) to the measurement of the physical properties of a nanomaterial, carbon nanotube.

In sciences, there are various ways to measure a dimension. The measuring instruments or methods employed depend on the scale at which metrology is approached. In order to describe the issues at stake for measurement at a given scale, we present the methods employed for the measurement processes at two scales of interest for scientists, namely the infinitely small, which corresponds to the Planck length of 1.6×10^{-35} m, and the infinitely large, which corresponds to the diameter of the Universe evaluated at 8.8×10^{26} m. This is to help the reader understand that, even though becoming an expert in a scientific field or in a given subject is not the objective, it is necessary to understand some basic tenets in order to master the methods used for successful metrology at a given scale.

In 1899, Planck determined a unit of length $l_P = \sqrt{(Gh/2\pi c^3)} \approx 1.6 \times 10^{-35}$ m, referred to as Planck length, based on fundamental constants: G, gravitational constant (6.6×10^{-11} Nm^2 Kg^{-2}), h, Planck's constant (6.64×10^{-34} Js) and c, the speed of light ($2.99,729,458 \times 10^8$ ms^{-1}). This length cannot be measured with the measurement technologies available on Earth. Indeed, the smallest length measurable at the LHC (Large Hadron Collider) of CERN, the particle accelerator in which two protons are made to frontally collide in a ring of 26,659 km, which led to the discovery in 2012 of the Higgs boson, is approximately 10^{-16} m, which is 19 orders of magnitude higher than the Planck length. CMS and ATLAS detectors were used in the observation of the Higgs boson, the latest prediction of the standard model not yet observed. The measurement at the scale of 10^{-16} m is made by compressing energy to reach an infinitely small spatial volume.

The principle of measurement at the scale of fundamental particles is mainly based on three relations: the de Broglie relation between the momentum p and the wavelength λ, $p=h/\lambda$, which introduces the wave–particle duality for matter; the relation that links the energy E of a particle to its wave frequency or wavelength λ, such as proposed by Einstein to explain the photoelectric effect $E = hc/\lambda$; and the relation that links the energy E of a particle of rest mass m to its rest mass energy and to its kinetic energy associated with its momentum $p=mv$, $E^2= m^2c^4 + p^2c^2$, as mentioned in Einstein's special theory of relativity. In the above formulas, v is the speed of the particle of mass m and c is the speed of light. The energy E can also be expressed by the formula $E= \gamma mc^2$, where γ is given by $\gamma =1/\sqrt{(1-v^2/c^2)}$. The speed of a particle is therefore given by $v/c=\sqrt{(1-(mc^2/E)^2}$.

In the LHC, the energy of a proton is 7 TeV ($1.2\ 10^{-6}$ J), far higher (by a factor of 7,500) than its rest energy, mc^2, which is 938 MeV. The formula for speed can then be rewritten as $v/c = (1-(m^2c^4/2E^2))$, which is equal to 1 to the nearest 10^{-8}. Using the relation $E= hc/\lambda$, the resulting value of the wavelength is of the order of 10^{-16} m, which gives the dimensions that can be reached in the LHC. The mass measured during two experiments at CERN in the LHC (8 TeV in 2012 and 13 TeV in 2015) is confirmed to the value of 125 GeV.

To detect the Higgs boson, a particle of mass 125 GeV associated with the Higgs field, while the mass of a proton is 938 MeV, the proton is accelerated and consequently its kinetic energy is increased so that its energy given by $E= \gamma mc^2$ significantly exceeds 938 MeV (8 TeV in 2012 and 13 TeV in 2015). The disintegration of colliding protons, each contributing an energy load of 8 TeV or 13 TeV, releases sufficient energy so that the Higgs boson can be expected to emerge during the recombination of subatomic particles. As the Higgs boson decays quasi-instantaneously after its emergence, the products of its decay must be analyzed to identify the excess energy and therefore the excess mass about 125 GeV.

It is worth noting that at the Planck length, the required energies that cannot be expected in a particle accelerator would lead to the emergence of black holes.

The opposite dimensional extreme towards the infinitely large corresponds to the spatial extent of the Universe, whose estimated value according to cosmologists is 10^{26} m. In cosmology, the observable Universe is a term used to describe the visible part of our Universe, the point from which light reaches us. It is a sphere whose limit is located at the cosmological horizon, having the Earth at its center. It is therefore a relative notion, as for other observers located somewhere else in the Universe, the observable sphere would not be the same (while its radius would be identical).

In cosmology, distances are measured in light-years. A light-year is the distance that light travels in one year, which corresponds to approximately 9.5×10^{12} m. The megaparsec, which is 3.26 million (3.26×10^6) light-years, is another unit of distance that is also specific to extragalactic astrophysics. Finding the size of the Universe involves accurate measurements of fossil radiation, or of the cosmic microwave background (CMB) radiation that originated in the Big Bang and can be used to determine the volume filled by the Universe since its creation. Predicted for the first time by Ralph Alpher in 1948 in his thesis work, CMB was discovered by Arno Penzias and Robert Wilson at "Bell Telephone Laboratories" during the development of a new radio receiver following the interferences detected independently of the orientation of the antenna they were building. While in a first approximation CMB is isotropic, accurate measurements of this radiation lead to determining H_0, the Hubble constant, which indicates the rate of expansion of the Universe.

In cosmology, detectors are above-ground telescopes. The WMAP (Wilkinson Microwave Anisotropy Probe) satellite launched in 2001 enabled the detection of CMB with good accuracy. Its intensity varies slightly in different directions of the sky and the fluctuations can be determined. Extremely accurate measurements of the WMAP in 2003 made it possible to calculate a value of H_0 of 70 kilometers per second and per megaparsec, which is within 5% in the hypothesis of a constant rate of expansion. Since the Universe is accelerating, during its expansion, the correction brought to H_0 made it possible to estimate the age of the Universe to 13.75 billion years, with a 0.1 billion margin of error. It is the scale fitting the domain to which corresponds the age of the Universe deduced from observations related to the Big Bang based on the inflationary model in an expanding Universe.

After the Big Bang, the elementary subatomic particles had no mass and could travel at the speed of light. After the expansion of the Universe and its cooling, the particles interacted with the Higgs field and consequently gained a mass.

In the history of the Universe, the elementary particles interacted with the Higgs field, 10^{-12} s after the Big Bang. The value of 125 GeV is considered as the critical value between a stable universe and a metastable universe. The "standard model of cosmology" elaborated at the beginning of this century, towards 2000, is probably at present the best model that enables the description of the evolution of the Universe, the significant stages in the history of the observable Universe as well as its current content, as revealed by astronomical observations. The standard model describes the Universe as an expanding homogeneous and isotropic space, on which large structures are overlaid as a result of the gravitational collapse of primordial inhomogeneities, which were formed during the inflation phase. There are still questions to be addressed, such as the nature of certain constituents of the Universe, black matter, and black energy and their relative abundance.

The inflationary model relies on the hypothesis of the Universe expanding with an exponential acceleration $R(t){=}R_0\exp(H(t)t)$, 10^{-30} s after the Big Bang, where $H(t)$ is the Hubble constant. This constant is measured from the Doppler effect, which explains the red shift of the light radiation emitted by a distant star that is receding from the point of observation. The inflationary model allows for a plausible interpretation of the CMB isotropy, with relative variations of the measured temperature of 10^{-5}. Based on the data provided by the Hubble, COBE (Cosmic Background Explorer) and WMAP (Wilkinson Microwave Anisotropy Probe) telescopes, as well as by the BOOMerang (Balloon Observations Of Millimetric Extragalactic Radiation ANd Geophysics) and MAXIMA (Millimeter Anisotropy eXperiment IMaging Array) experiments, scientists were able to determine the age of the Universe is 13.75 billion light-years.

The Universe is presently in accelerated expansion: if its speed is 70 km/s at 1 Megaparsec, it doubles at 2 Megaparsec, reaching 140 km/s and so on. Considering the Doppler shift or the red shift for the receding stars, and therefore the fact that not only are the stars receding, but also those that are twice farther recede twice faster, therefore considering the metrics applicable to the space that is stretching while galaxies are receding, the 13.8 billion years between the beginning of the rapid expansion of the Universe 10^{-30} s after the Big Bang amount to 46.5 billion light-years, which is a radius of 93 billion light-years. Obviously, the light of stars that are at the periphery or at the cosmological horizon can no longer reach us, but as what we observe today goes back to the time needed for light to reach us while traveling a distance in a stretching space.

These two examples show that at each dimensional scale, besides the appropriate experimental measurement techniques required for observation, we must have a good mastery of the theories adapted for the interpretation and analysis of the gathered data. At each scale, the engineer must acquire specific knowledge elaborated in the laboratories and develop the competences to enable the mastery of technologies and the implementation of innovations.

This book which provides applications for Reliability of Multiphysical Systems Set Book 2 (*Nanometer-scale Defect Detection Using Polarized Light*), focuses on knowledge elaborated at the nanometer scale for applications in the field of engineering sciences. The subjects approached are related to simulation experiments and engineering of nanometer-scale systems. The light–matter interaction has a special place among the subjects addressed, because the analysis of the properties and characteristics of matter is most often possible due to light being used as a probe. Similarly, simulation according to theoretical models based on quantum mechanics principles requiring field theory is also given particular attention.

Nanotechnologies and nanosciences are identified as sources of breakthrough innovations that will lead to the development of technologies that are considered primordial in the 21st Century. They should be deployed in eco-innovations and will increasingly become pervasive in the societal applications in various sectors. Without pretending to provide an exhaustive list, several examples are worth being mentioned: new energies and their recovery and storage, water purification, new materials that are lighter and more resilient for land and space transportation, construction and buildings, information technologies with quantum computers, embedded electronic systems and factory 4.0. The trend according to which states throughout the world offer financial support for the development of long-term projects in this field dates back to the beginning of the 21st Century. This is a reflection of the economic growth potential in nanotechnologies.

Similar to the inflationary model proposed by cosmologists to explain the countless galaxies and planetary systems, suns and black holes that constitute them, there was also a sharp increase in the volume of activities in nanosciences. The subjects approached in this book and in the Reliability of Multiphysical Systems Set Book 2 (*Nanometer-scale Defect Detection Using Polarized Light*) concern the field of engineers working in mechatronics, robotics and computation in modeling and simulation, for the societal spin-offs of nanotechnologies in the fields of land and space transportation, handicap, information and simulation technologies in a systemic approach. The level of knowledge acquired by the engineer should make innovation in nanotechnologies possible.

The contents of *Nanometer-scale Defect Detection Using Polarized Light* and *Applications and Metrology at Nanometer Scale 1 & 2*, jointly written by three authors, aim to develop knowledge that is essential at the nanometer scale, enabling trainee-engineers or engineers to develop nanotechnology-based devices or systems. To promote the deployment of nanotechnologies, the authors of these three books whose joint competences and experiences associate know-how in fundamental physics, engineering sciences and industrial activities cover a wide spectrum of application domains. *Nanometer-scale Defect Detection Using Polarized Light* builds a theoretical and experimental basis for understanding nanometer-scale metrology. This book in two volumes, *Applications and Metrology at Nanometer Scale*, enriches this theoretical basis with applications in the form of corrected exercises.

1

Nanometer Scale

The methodologies implemented for the elaboration and analysis of materials or objects on the nanometer scale are part of nanotechnologies. The fundamental research conducted at this scale, which is referred to as nanoscience, generates knowledge on the innovative properties associated with nanometer dimensions, as well as on the relevant methods of analysis. Nanotechnologies concern technologies used for the industrial manufacturing of devices carrying a commercial and societal value, while following the safety limits applicable to nano-objects manipulation and use. It is essentially multidisciplinary and relies on chemistry, physics, materials science, simulation, information and communication technologies, engineering science and many other disciplines. Mechatronics and robotics for terrestrial and spatial transportations and the technologies for connected objects are concerned by the developments in nanotechnologies. Working at the nanometer scale requires especially adapted metrology, as measurements of the properties of nanodevices often lead to relative values whose standard deviation is of the same order of magnitude as the measured value. Indeed, in most cases, to qualify the measured values, simulations should be conducted in parallel with the measuring process.

1.1. Introduction

Historically, the landmark of the foundation of nanotechnology is the lecture given by R. Feynman, recipient of Nobel Prize for Physics in 1965 for his contribution to quantum electrodynamics. The lecture was entitled "There is plenty of room at the bottom", and was given on December 29, 1959, at the annual meeting of the American Physical Society at California

Institute of Technology. On this occasion, he stated: "The principles of physics, as far as I can see, are not against the possibility of manipulating things atom by atom".

The term "nanotechnology" was coined by N. Taniguchi in 1974 [TAN 74], to describe the processes of thin-film deposition during the elaboration of "wafers" of semiconductors. The scanning electron microscope developed in 1981 [BIN 82a, BIN 82b, BIN 83 BIN 86a], for whose invention G. Binnig and H. Roher received the Nobel Prize for Physics in 1986, can be used to visualize surfaces at atomic or nanometer scale. In retrospect, some sort of Gant chart can be drawn to quantify the time lapse of over two decades between Feynman's idea in 1959 on atom manipulation, the elaboration in the 1970s of materials or nano-objects structured at nanometer scale, including the invention of the term nanotechnology in 1974 and, finally, the development of the instrument enabling the effective visualization and manipulation of the atom, in 1981. Two major fields of investigation corresponding to different temporalities can be identified: a first phase for the controlled synthesis or elaboration of nano-objects, and a second phase for measurement and analysis. The acquisition of knowledge used by the engineer to implement innovations requires the development of imaging tools and especially of nanometer-scale analysis methods. The efficiency of the overall scientific effort depends on the availability of multidisciplinary competences, which involves a systemic approach.

Some of the most fascinating problems in all science fields involve multiple temporal or spatial scales (cosmology, materials science, elementary particles, etc.) so that in the science field, the scales of interest for scientists range from Planck length of 10^{-35} m to the distance of 10^{24} m, which, in cosmology, corresponds to the diameter of the volume of observation (local group) of the Universe since the Big Bang.

In the field of technologies and engineering sciences, the invention of the transistor in the 1950s by J. Bardeen and W. Brattain [BAR 48] and W. Shockley [SHO 49] of Bell Laboratories opened the way to micro-technologies. The integration density of transistors consequently doubled every two years, in accordance with Moore's law [MOO 65]. As shown in Figure 1.1, the following stage of technology development is in the field of nanotechnologies, a source of innovations for the 21st-Century industry.

Figure 1.1 gives an overview of the various scales (on the X axis) that should be considered in materials sciences and in scientific subjects (quantum physics, chemistry, materials sciences and engineering) that elaborate the knowledge in the considered field (nanoscopic, microscopic, mesoscopic, macroscopic). The scale of interest for technological applications is naturally the macroscopic scale, the dimension perceived at human scale in society. The relations between material structure and properties determine the behavior of the material at the macroscopic scale. Figure 1.1 also indicates the properties or phenomena studied at various scales by simulation, providing a non-exhaustive list of several theoretical models developed for apprehending the behavior of the material (*ab initio* calculations, molecular dynamics, Monte Carlo simulation, dislocation dynamics and Monte Carlo dynamics, polycrystalline models, finite element method).

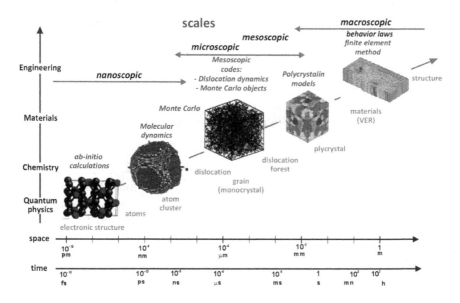

Figure 1.1. *Various spatial and temporal scales in materials sciences. For a color version of this figure, see www.iste.co.uk/dahoo/metrology1.zip*

The term nano originates in the Greek term "nannos" or "ναννος", which means "excessively short or dwarf". In metrology, which is the science of measurement, a nanometer corresponds to 10^{-9} m. While human hair size of 10^{-4} m is small on the meter scale, it is huge at the nanometer scale. It is the scale corresponding to molecules. It was agreed to classify any object whose size ranges between 10^{-10} m and 10^{-7} m as belonging to the nanometer scale, though the term nano more generally describes measured dimensions that range between 1 nm and 100 nm. This interval contains, for example, dimensions corresponding to the size of viruses (100 nm), deoxyribonucleic acid or DNA (10 nm) and molecular structures (1 nm). The term nano generated a large number of terms whose prefix is nano. The terms employed in this book are mainly nanoscale, nanometric, nanosciences and nanotechnologies, thus referring the reader to works specifically dedicated to nanosciences and nanotechnologies.

Nanoresearch is conducted in many domains: nanoelectronics, modeling, simulation and design of nano-objects, nanobiotechnologies, nanomedicine, nanophotonics, nano-characterization, nanometrology, nanolithography, nano-fabrication of nano-objects, nanotoxicology in health, environment and nano-security fields. The knowledge and know-how related to all these domains can be found in numerous works that were published after the emergence of nanoscience in the 2000s. The readers who are interested in gaining in-depth knowledge in a specific field are referred to [RAT 02, LAH 04, DUP 06, LAH 06, LAH 07, LAH 10, LOU 16, KNE 20, LEE 20]. Being in continuous and rapid evolution in all the disciplines since 2000, nanoscience and nanotechnologies are the object of many calls for projects throughout the world.

On the nanometer scale, the classical theories of Newtonian mechanics are not adapted and the principles of quantum mechanics should be applied for the interpretation of the mechanics of physics and chemical phenomena. In classical mechanics, distant interactions are modeled by forces that act instantaneously throughout the space. While possible according to classical theory, simultaneous measurement of position and speed is not possible in quantum mechanics, following Heisenberg's uncertainty principle $\Delta x \Delta p \approx h/2\pi$, which must hold.

On the other hand, Maxwell's theory of the electromagnetic field, which is built in terms of electric and magnetic fields, is in agreement with the quantum description for matter–light interaction, as shown in [DAH 16].

In his works on the attraction of spheroids, Laplace shows that "the calculation with partial differences" can be used to locally formulate the interaction between two masses in terms of potentials. Laplace writes on page 358 [LAP 94] that "a sphere attracts an external point as if all its mass were united in its center" and on page 382 [LAP 94] that "In the planetary theory, the objective is not to determine their equilibrium in absolute space, but only the equilibrium of all their parts around their centers of gravity". Nevertheless, if gravitation is formulated in terms of fields or potential, as Laplace did in 1785 [LAP 94] to locally describe gravitational force, classical mechanics may be regarded as an approximation of the general theory of relativity. Unlike Maxwell formalism, which describes the electromagnetic phenomena in terms of electric and magnetic fields, it cannot be directly inserted in the quantum theory.

The transition from classical mechanics to QM is possible through the principle of correspondence between Poisson brackets defined in Hamiltonian mechanics and the commutators between operators acting in Hilbert space according to Heisenberg formalism. The formulation of the theory of gravitation in terms of gravitational fields as used in general relativity should be compared to the quantum field theory, which is more appropriate for the description of interactions between elementary particles.

All these considerations show that different formalisms were developed in order to describe the physical phenomena on various scales. To innovate in the field of nanotechnologies, the engineer should therefore apprehend and master the physical concepts that are currently being developed in research laboratories.

On the nanometer scale, implementing a technique for the measurement of a physical property is not a simple task, and when a measurement is made, its repeatability is not guaranteed. In general, in the absence of a standard as a reference, accuracy is hardly reached. And if precision is achievable, it requires a lot of time to be set up at the nanometer scale. Consequently, there are few measurements really achieved and the implementation of measurement instruments is not an easy task. Furthermore, measurements should also rely on theoretical models.

Experimental and theoretical examples for the study of materials at the nanometer scale are given in Chapter 6 of [DAH 16]. The first chapter of [DAH 16] provides an overview of the elaboration, characterization and analysis techniques relevant to nanotechnologies. There are two major fields of investigation in nanotechnologies, namely the synthesis or elaboration of nano-objects while controlling the manufacturing process and the process of characterization and analysis, which involves the development of imaging and analysis tools at the nanometer scale. These studies rely on quantum mechanics principles.

Based on research works conducted in the laboratories of the University of Versailles as part of fundamental research [DAH 04a, DAH 04, FRA 05, HAM 06, HAM 07, NOU 07] or in relation with the industrial environment, three examples are considered to illustrate three techniques for the elaboration of nanomaterials samples: spin coating, cathode sputtering and laser ablation. Similarly, four techniques for sample study and characterization are presented: transmission electron microscope (TEM), scanning electron microscope (SEM), atomic force microscope (AFM) and ATR spectroscopy.

1.2. Sample elaboration

The samples studied in nanosciences are manufactured either in bulk form or as a thin film deposited on a substrate by chemical, physical and chemical or physic-chemical processes. A sample is classified as a nanomaterial if at least one of its dimensions ranges between 1 and 100 nm.

A thin film always involves a substrate, irrespective of the procedure employed for its manufacture (it may happen that a thin film is separated from its substrate). Consequently, the substrate has a very strong influence on the structural properties of the deposited film. The nature of the substrate influences the physical properties of a mono-material thin film, all other characteristics being the same. Thus, the physical properties of a thin film are slightly different depending on whether it is deposited on an amorphous insulating substrate, such as glass, or on a silicon monocrystalline substrate, for example. It is worth noting that a thin film is anisotropic by construction.

There are numerous methods for preparing thin films. Only those most commonly used in the field of electronics and technology in relation to transport, optics, magneto-optics and heat properties are mentioned here. The main manufacturing methodologies used by the manufacturers of active or passive electronic components are based on physical processes for the deposition of material on a substrate. Therefore, the thin film increases in thickness from zero and this growth must be controlled.

In the case of chemical deposition, the films grow through a chemical reaction between a substrate maintained at an optimal deposition temperature in the presence of an inert gas in a low pressure preparation chamber and a precursor fluid, which results in a solid film that covers the whole surface. Depending on the precursor phase, chemical deposition is differently classified. Some examples of coating are spin coating, dip coating, atomic layer deposition (ALD), electro-deposition and, finally, chemical vapor deposition (CVD), and the variants associated with CVD differ depending on the process used to initiate the chemical reactions and on the process conditions, such as plasma-enhanced (PECVD), metallo-organic (MOCVD) and very low-pressure CVD.

In the physical deposition processes, the material to be deposited is generated in solid, liquid or gaseous state, and then it is physically carried on the surface of the substrate. This stage takes place in a preparation chamber under vacuum. The processes through which the matter to be deposited is generated are essentially mechanical, thermal or electromechanical, such as thermal evaporation, sputtering or ion plating. The choice of the deposition process depends on a certain number of factors, such as the substrate structure, the deposition temperature, the deposition rate and the source of material. The deposited films can be structured by other techniques, such as photolithography.

Two large families of physical methods can be distinguished in practice. One of them involves a carrier gas that moves the material to be deposited from a recipient to the substrate and is related to scattering techniques used in the manufacture of active components. The other one involves a very low-pressure environment and the material to be deposited is carried by an initial thermal or mechanical pulse. To elaborate a thin film, we must choose the material to be deposited, the substrate and the deposition technique.

The three most common thin-film deposition techniques are cathode sputtering, molecular beam epitaxy and laser ablation.

– *Cathode sputtering*. A target is bombarded with an inert gas (argon) in order to sputter the atoms in the target. Applying a potential difference between the target (cathode) and the substrate (anode), the sputtered atoms are accelerated and deposited on the substrate.

– Molecular beam epitaxy. Conducted under high vacuum, this method involves the evaporation of the material to be deposited by heating, radiation or electron bombardment.

– LASER ablation. Groups of atoms in the target are "evaporated" by means of a high-fluence LASER beam.

Unlike magnetron or diode sputtering, which relies on a mechanical principle of atomic or more precisely ion bombardment, vacuum or ultra-high vacuum evaporation relies on a thermal principle. Heat brings matter to its melting point and then to its evaporation point.

Vacuum evaporation relies on two elementary processes, namely the evaporation of a heated source followed by condensation to solid state of the matter evaporated on the substrate. Matter heating can be generated by several techniques leading to its evaporation: Joule effect, through which a current of normally several hundred Amperes passes through the matter to evaporate, electron bombardment (evaporation using an electron gun), effusion evaporation, magnetron cathode sputtering, diode cathode sputtering and laser ablation evaporation.

Molecular beam epitaxy (MBE) is conducted under ultra-high vacuum. It involves the evaporation of the material to be deposited by heating, radiation or electron bombardment on a substrate. With this technique, one or several molecular beams of atoms can be directed towards the substrate to achieve epitaxial growth. It enables the growth of nano-structured samples of several cm^2 at a speed of about one atomic monolayer per second. An MBE setup consists of an introduction airlock equipped with a turbomolecular pump and a preparation chamber containing:

– a gun enabling ion bombardment of the surfaces (Ar+ ions of energy ranging between 1 and 8 keV);

– a furnace enabling heating up to 1,800°C by combining Joule effect up to 500°C and electron bombardment for temperatures beyond;

– a micro-leakage valve to introduce gas in the chamber. The temperature during growth is measured by a thermocouple and/or an infrared thermometer.

To illustrate the thin-film deposition techniques at nanometer scale, descriptions of spin coating, cathode sputtering and laser ablation techniques are detailed below.

1.2.1. *Physical and chemical method: spin coating*

One method for chemically preparing a polymer sample in the form of thin film on a substrate at nanoscale is spin coating. It is worth noting that there are several methods for preparing thin films, depending on the nature of the polymer, thickness and application, namely spin coating, dip coating or serigraphy.

This section presents the process used in a thesis work at the University of Versailles St Quentin, in a joint project with the transportation industry, funded within MOVEO, a competitiveness cluster [KHE 14]. The embedded electronics systems from car suppliers in the automotive industry comprise plastic materials, notably for coating electronic components. To ensure the reliability of the encapsulated devices, the quality of the plastic cases of the electronic components must be controlled. Due to their low costs and high performances, electronic cases are essentially made of plastic materials. In this respect, plastic cases are most often made from epoxy resin and silica. Water absorption is the first cause of deterioration of electronic cases, resulting in the formation of cracks and delaminations of the coating. Aging studies in humid or aqueous environments contribute to a better understanding of the various paths corresponding to physical or chemical degradations of resins. The presence of water more or less related to the polymer is identified as a main cause.

The study of interfaces at the nanometer scale was conducted on silicone gels used for the encapsulation of electronic circuits. These polysiloxane-based gels are polymers characterized by a silicon–carbon Si–C bond and by a silicon–oxygen Si–O bond. They are very stable at high temperatures $\geq 180°C$ and provide an electric insulation of 20 kV/mm, being both resistant to chemical products and moisture proof. The silica surface is characterized by the groups: silanols (Si–OH) and siloxanes (Si–O–Si). Interactions between the reactive functions of the polymer and the silanols of the silica surface are susceptible to appear at the silica surface. On the other hand, the presence of hydroxyl functions on the polymer chain supports the absorption and the fixation of the polymer by hydrogen bond to the surface of the silica particle. A coupling agent is added to modify the silica surface in order to favor the bonds between the organic polymer and the charges. It generally takes the form of an alkoxysilane. During the cross-linkage reaction, this agent forms stable chemical bonds between the polymer and the charges. One of the coupling agents that is currently used in the industry is glycidyloxypropyltrimethoxysilane (GOTMS). The coupling agent can also be very useful at the interface between metal and resin, as it forms a hydrophobic film at the surface of the metal and enables a very good adhesion between the metal and the resin. The studied polymers were either single-component silicone gels or two-component silicone gels. The characterization of samples by ellipsometry at nanometer scale was presented in Chapter 8 of [DAH 16].

Spin coating is widely used for the deposition of polymer films on electronic components. It involves the deposition by a syringe of a given volume of polymer solution on the surface of a substrate. A vacuum pump enables the fixation by aspiration of the substrate on a rotary support during the rotation for the efficient spread of the film and a mechanical stability of the whole even at 10,000 rpm. The device spinner can reach a rotational speed of 10,000 rpm and an acceleration/deceleration from 1 to 7,500 rpm. The various stages are schematically represented in Figure 1.2.

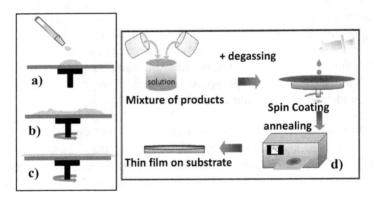

Figure 1.2. *The stages of spin coating technique: a) deposition of volume, b) spreading by rotary spinner, c) homogeneous thin film, d) spin-coating-based manufacturing process. For a color version of this figure, see www.iste.co.uk/dahoo/ metrology1.zip*

By spin coating technology, roughness and thickness can be kept under control, their values ranging from about 100 nanometers to several micrometers. Among other factors, the final thickness depends on the concentration of the polymer solution (viscosity) and on the rotational speed (centrifugal force) of the device. The thickness d of a thin film deposited by spin coating is indirectly proportional to the angular speed ω of the spinning plate. It can be written that $d \approx \omega^{-n}$, where the exponent n depends on the solvent. In the absence of evaporation, d varies with time t and speed, such that $d \approx \omega^{-1}t^{-1/2}$ and if the evaporation rate is constant, $d \approx \omega^{-2/3}$. However, it is observed that $d \approx \omega^{-1/2}$. A calibration of the device is required to determine the volume of solution as a function of the desired thickness given the viscosity of the product.

An accurate protocol must be followed before depositing the solutions on the chosen substrate. Quartz (SiO_2) is generally used as a model substrate before trying other substrates that can otherwise be made from ceramics, metal (Al), semiconductor (Si, Ge) or alloy (Cu/Ni, Cu/Ag). Substrates must first be cleaned. If they are not prone to oxidation, an ultrasonic bath is used: the substrate is immersed in water, respecting the allowed volumes between a minimal and a maximal quantity, the bath temperature (T_C: maximum 45°C) and the required time.

There are three stages in a sample preparation: cleaning of the substrate, preparation of the polymer solution to be deposited, then annealing of the polymer.

Surface treatment is essential for avoiding the presence of impurities on the substrate surface. The slightest impurity can generate a contamination that leads to the separation of deposited layers.

– To clean a silicon substrate, an acetone solution at 30°C is used as the ultrasonic bath. After 15 minutes, the wafer is removed from the bath and abundantly rinsed with distilled water followed by ethanol and finally dried by a flow of pressurized dry air. The wafer is then ready to be used as the substrate for the polymer film.

– For a Cu/Ni alloy, the cleaning is done with acetone and distilled water to prevent damage of the nickel layer.

– Quartz or metal (Al) substrates are submitted to ultrasonic cleaning with distilled water and acetone. Each of these baths is applied for 15 min. The substrate is then dried under a flow of pressurized dry air.

Polymer solutions are prepared using a double-barrel syringe with several spare parts. Before the deposition of polymer solutions, it is cleaned by solvents, such as acetone or ethanol. Polymers are generally provided in liquid form of variable viscosity. When the polymer is composed of two components, namely a resin referred to as R and a catalyst referred to as C, R and C must be mixed in a proportion according to the provider specifications (1:1 or 10:1). Since the mixture must be homogeneous, a magnetic stirrer is used for 2 min at 200 rpm for the mixing. Then, several drops are deposited on the substrate using a measuring syringe, before starting the above-described spin-coating process. Programming of the processing steps is set beforehand, depending on the calibration curve of the device. For current polymers, the standard deposition parameters are 1,500 rpm for 55 s and a start ramp of 55 rpm per second. The same deposition conditions are generally maintained for various preparations. The preparation setup is placed under a 30 mbar vacuum for degassing, to prevent bubble formation. This step is carried at ambient temperature.

The next stage after film formation is the curing at a given temperature depending on the polymer to be cross-linked. The cross-linkage mode and the polymer temperature are found in the provider datasheet. If the provided values are respected, a fully polymerized sample can be obtained. In general, if the recommended temperature is exceeded, the material irreversibly deteriorates. Cross-linkage requires a furnace that can heat up to 2,000°C. A degassing chamber comprising a vane pump for low vacuum prevents in a gray room the presence of ambient impurities in the form of dust or aerosol present in the air.

1.2.2. *Physical method: cathode sputtering*

Cathode sputtering (Figure 1.3) involves the bombardment of a target with an inert gas, argon maintained under a reduced pressure ranging between 5×10^{-1} and 10^{-3} Torr, in order to sputter the atoms in the target. It is a mechanical process that depends on the momentum lost by the collider when the Ar^+ ion collides with the atom in the bombarded target. This process does not depend on temperature. Applying a potential difference between the target (cathode) and the substrate (anode), the sputtered atoms are accelerated and deposit on the substrate. This technique is also known as "sputtering deposition" adequate for thicknesses below 3 μm.

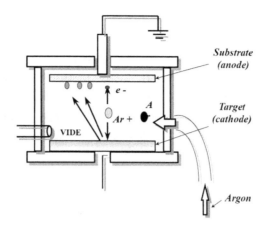

Figure 1.3. *Cathode sputtering. For a color version of this figure, see www.iste.co.uk/dahoo/metrology1.zip*

This technique was used to prepare multi-layers of SiO_2 and TiO_2 films in order to create a 1D magneto-photonic crystal. The target material to be deposited is introduced in a vacuum chamber in the form of a plate of several millimeters thickness and whose dimensions are equal to those of the plate to be coated. This target is fixed at a cathode that is cooled and is maintained under a negative voltage ranging between 3 kV and 5 kV. The anode, which plays the role of the substrate holder, is set in parallel to the target, at a distance of several millimeters. It is generally grounded, for ease of use. For a residual pressure in the chamber maintained between 1 and 10^2 Pa, the electric field between the two electrodes generates the ionization of the visible residual gas in the form of a luminescent plasma cloud. An electric current is then established between the two electrodes, through the conductive plasma formed of electrons, which are attracted by the anode, and positive ions, which are attracted by the target cathode.

A 1D magneto-photonic crystal is a multi-layer stack of a dielectric and a magnetic material characterized by enhanced magneto-photonic properties. The magneto-optical response (Kerr rotation of polarized light) is studied as a function of the thickness of the films by calculation prior to the crystal manufacture and its characterization. The deposition parameters having been calculated, alternating thin films of SiO_2 and TiO_2 are deposited by cathode sputtering according to the calculated specifications. The criteria to be met are:

– state of the surface: the stacks of dielectric films require the least possible individual roughness of the films to minimize the perturbation generated by this type of defect in the developed structures;

– nature of the material: there should be a significant contrast between the materials to be stacked, with the lowest possible extinction coefficient, which explains the choice of TiO_2 and SiO_2;

– deposition speed: since the development of multilayers requires a significant number of periods, high deposition speeds of single layers are chosen for the stacks over a reasonable period of time;

– deposition temperature: it should be the lowest possible to limit inter-diffusion at the interfaces between SiO_2 and TiO_2 layers.

AFM is used for the study of surface roughness, and ellipsometry is used to determine multilayer thickness and (optical n and k) indices.

The 1D magneto-photonic crystal is thus manufactured using cathode sputtering for multi-layer stacking [HAM 06, HAM 07] and is studied for its properties as smart material at the nanometer scale.

1.2.3. *Physical method: laser ablation*

The principle of thin-film deposition by laser ablation (Pulsed Laser Deposition) involves a nanosecond or femtosecond pulsed laser beam focused on a bulk target (oxide target), in an ultra-high vacuum chamber. The first pulsed laser depositions (PLDs) were conducted in 1965 by Smith and Turner [SMI 65], long before it became widespread, thanks to the development of high-density short-pulse lasers initiated in 1987 [DIJ 87].

The description which is given as an illustration of laser ablation technique concerns the research works conducted at the University of Versailles St Quentin in a laboratory of the National Center for Scientific Research in France for the study of thin films of magnetic oxides of nanometric thickness [DAH 04a, DAH 04, FRA 05, HAM 06, HAM 07, NOU 07]. Two power UV lasers are used: a chemical excimer laser (248 nm) and a solid laser with doubling and tripling crystals. Beyond a certain fluence (amount of energy per unit area or surface energy density), a light plume can be noted on top of the target. This plume corresponds to a directed expulsion of matter, in the form of a plasma, which can be described as a "gas" of ions, atoms and electrons. These particles are ejected at a high speed while the whole preserves its electrical neutrality. This ejected matter is then collected on a substrate, which is generally a quartz plate located above the target and maintained at a certain deposition temperature. Figure 1.4 shows a schematic representation of the setup corresponding to the laser ablation system.

The samples are prepared in a very high-vacuum chamber under a pressure above 10^{-6} Torr. A light beam from a solid pulsed laser of 10 ns that operates in UV at 385 nm with a repetition rate of 10 Hz is focused on a target that is generally prepared by sintering in the chamber. A reflecting device composed of mirrors that can randomly scan the laser beam on target thanks to an algorithm-based control of the device mirrors is used. The diameter of the laser beam transmitted through a silica window randomly describes a square on the target. Fluence ranges between 2 J/cm^2 and 20 J/cm^2. The substrates (quartz, MgO, etc.) located at the required distance (40 mm to 70 mm) from the target are placed on a holder maintained at an optimal controlled deposition temperature (maximal value of 1,000 ± 5°C) in an oxygen atmosphere under a pressure of up to 500 mTorr.

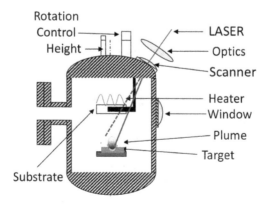

Figure 1.4. *Schematic representation of laser ablation deposition setup. For a color version of this figure, see www.iste.co.uk/dahoo/metrology1.zip*

The solid laser is a flash-pumped Nd: YAG laser. At the output of the laser cavity, a frequency-doubling KDP crystal is used to generate the second harmonic of laser light at 532 nm. A crystal mixing radiation at 1,064 nm and 532 nm is used to obtain a beam of UV light at 355 nm. Concerning the value of laser radiation fluence at the level of the material, laser radiation–material interaction studies show that it can be classified into three categories:

– Low fluence: the energy deposited by the laser diffuses in the material. It can activate chemical processes at the surface of and/or within the material. The vaporization rate is negligible.

– Intermediate fluence: laser energy is at equilibrium with the losses by heat diffusion and by fusion and vaporization processes. Vaporization is more significant and vapor remains transparent to radiation.

– High fluence: under this regime, a plume appears above the surface of the material, which is generally no longer transparent to laser radiation. Moreover, it may react with the ambient atmosphere and the target material.

The existence of "threshold fluences" beyond which these various phenomena occur can be experimentally observed. They depend on several parameters: radiation wavelength, nature and pressure of the ambient atmosphere and nature and state of the target surface.

A range of parameters can be varied during deposition:

– characteristics of the laser beam: power, beamwidth, duration of the pulse (between 5 and 10 ns for the pulse time at mid-height);

– pressure: between 10^{-4} and 10^{-6} mB;

– temperature of the ceramic heater: between 300 and 800°C;

– oxygen pressure: to maintain the stoichiometry of the sample and confine the plasma. It varies between 60 and 250 mTorr;

– the target–substrate distance: from 40 to 70 mm.

Since the way atoms or molecules are arranged at the surface of a layer determines its morphology and structure, this initial organization plays an essential role in its physical and/or chemical functionalities. Layer growth and morphology optimization is therefore decisive. The growth of layers obtained by laser ablation is achieved by two-dimensional nucleation of aggregates. At a low fluence, these aggregates are small and, in parallel with the subsequent formation of islands, atoms and aggregates scatter at the surface of the substrate, leading to the formation of layers on top of which aggregates grow. The morphology and structure of layers resulting from this growth mode depend on the conditions of deposition such as the target–substrate geometry, substrate characteristics (surface, nature, temperature), oxygen pressure or the fluence of the ablation laser. Layer growth control involves the optimization of these parameters by in situ controls. As the nature and quality of the deposited layer depend on many parameters (laser energy or fluence, the pressure of the residual gas in the chamber, substrate temperature, etc.), in order to control in situ growth, during deposition, a spectroscopic ellipsometer, as schematically represented in Figure 1.5, was set up in parallel with the preparation system [DAH 03, NOU 07, DAH 11, COA 12]. This non-destructive technique described in Chapter 8 of the [DAH 16] is used to determine layer thickness and optical parameters, i.e. those data required for the analysis of the measurements conducted in magneto-optical spectroscopy.

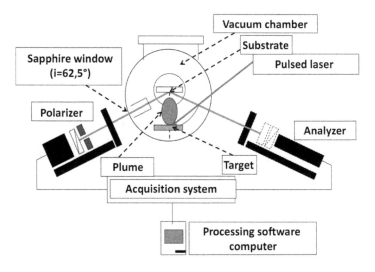

Figure 1.5. *Ellipsometry-based in situ control of thin-film growth. For a color version of this figure, see www.iste.co.uk/dahoo/metrology1.zip*

The results obtained on a thin film of perovskite structure magnetic oxide $SmFeO_3$ deposited on a quartz substrate are given in Figure 1.6(a) and (b) for various temperatures. The ellipsometry parameters $\cos(\Delta)$ and $\tan(\Psi)$ (Chapter 8 of [DAH 16]) are given depending on the wavelength for various temperatures ranging between 300 K and 1,095 K.

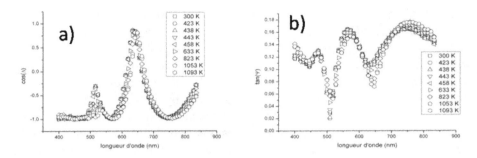

Figure 1.6. *Variation with temperature of a) cos (Δ) and b) tan (Ψ)*

The analysis of the curves shows that beyond 600 nm, cos (Δ) oscillates between ±1, which corresponds to a very low absorption zone. In the absence of thin film, the substrate gives a value of +1 for cos(Δ). When the thin film is deposited, cos(Δ) varies between -1 and +1, the number of oscillations increasing with the thickness of the film. Below 600 nm, in the absorption zone, the maxima remain below 0.1. When temperature increases, varying positions and amplitudes of the extrema can be noted. They shift towards longer wavelengths, which corresponds to an increase in the thickness of the film following the material growth. The amplitudes decrease significantly in the absorption zone, which corresponds to an increase in absorption with temperature. Concerning the parameter tan (Ψ), in the absence of thin film, the quartz substrate gives a ratio of the amplitudes r_p/r_s whose variation is linear in the spectral range from 300 nm to 850 nm, between 0.18 and 0.20. When the film is between the substrate and the air, a variation of tan (Ψ) around this straight line can be noted, with the extrema shifting towards long wavelengths when the thickness increases or the temperature rises. The excursion of the amplitudes, on either side of this line, is a function of thicknesses and absorption.

For the fabrication of a one-dimensional magneto-photonic crystal [HAM 06, HAM 07], the optical properties of TiO_2 and SiO_2 could be determined in the chamber fitted with the ellipsometer, as shown in Figure 1.7. In this case, the target being made of TiO_2, then TiO_2 deposited on SiO_2 could be characterized. The deposition was achieved using an excimer laser that emits a radiation of 248 nm during the discharge triggered by a thyratron in a gaseous mixture of krypton and fluorine. The energy of each pulse, with a duration of about 25 ns, varied over the range from 150 to 650 mJ; fluence was fixed at 1.8 J/cm^2. Laser repetition rate could be adjusted between 1 Hz and 20 Hz and was fixed at 10 Hz. Substrate temperature was maintained at 300°C, and the pressure of the oxygen in the chamber was measured to be 0.0373 mbar. It is worth noting that laser ablation of SiO$_2$ at this wavelength is not possible due to an absorption band at 5.01 eV in resonance with the frequency of the excimer laser so that the deposited energy is mainly dissipated by radiative relaxation that manifests as photoluminescence (Figure 6.8 in Chapter 6 of [DAH 16]).

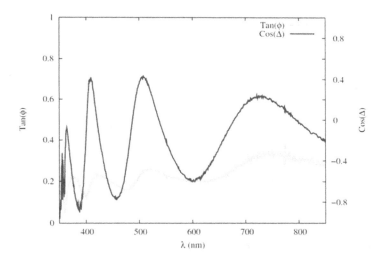

Figure 1.7. *Ellipsometric parameters of a film of TiO$_2$ on SiO$_2$ by laser ablation [HAM 07]. For a color version of this figure, see www.iste.co.uk/dahoo/metrology1.zip*

RHEED (reflection high-energy electron diffraction) [RUS 33, COL 72] is generally used as in situ probe to control the growth of thin films. This technique is sensitive to changes at the surface, due to the changes in structure or to adsorption. It involves the bombardment of a target surface with very high-energy electrons. Electrons are then diffracted in various directions with different intensities. The impact of electrons on a phosphor screen gives an image formed of bright spots that correspond to the arrangements of atoms on the surface. A beam of accelerated electrons (5–100 keV) under grazing incidence (0.5 to < 3 deg) is reflected on a surface. Due to their energy, the penetration of electrons is significant. But because of the grazing incidence, only few atom layers are probed. Hence, this technique is sensitive to surfaces.

Two types of diffraction are possible: by reflection and by transmission–reflection: 1) diffusion by a 3D crystal island (image in the upper part of the screen); 2) 2D surface diffusion by a plane surface (image in the lower part of the screen). The energy of electrons in a RHEED experiment may vary between 10 keV and 100 keV, but most of the instruments use energy between 10 keV and 35 keV. Let us recall that 1 eV is the kinetic energy gained by an electron that is accelerated under a potential difference of 1 V. The kinetic energy is $E_c = 1/2 \ mv^2$, where v is the speed of the electron and the electric energy is expressed in eV. The speed is $v = \sqrt{(2eV/m)}$ and using the momentum $p = mv$ and the de Broglie relation $p = h/\lambda$, it can be deduced that $\lambda = h/\sqrt{(2meV)} = \lambda_c$ for a non-relativistic electron. Under relativistic conditions, when energies are high, mass correction must be taken into account, and therefore $\lambda = \lambda_c(1/\sqrt{(1 + eV/2mc^2)}$. The corresponding wavelength varies between 0.0037 nm and 0.0146 nm for energies of 100 keV and 10 keV.

1.3. Characterization of samples

A thin film of a given material is an amount of that material having one of its dimensions, referred to as thickness, significantly reduced. This thickness (in nanometers) corresponds to the distance between two boundary surfaces (quasi bi-dimensionality). This 2D structure results in a perturbation of most of its physical properties. The essential difference with the bulk state is that the role played by the boundaries in the bulk properties is generally considered negligible. On the other hand, in a thin film, the effects related to boundary surfaces are predominant. The smaller the thickness, the more pronounced is this bi-dimensionality effect. Conversely, beyond a certain threshold, the effect of thickness becomes negligible and the material regains its properties of the bulk material. Any method of analysis requires a probe (electromagnetic radiation, a beam of energetic particles, a field, a mechanical penetrator, etc.) acting on a sample. The result of the interaction between this probe and the sample can be another radiation, particles, field variation, detected by a chain of measurement or by a behavior law.

To observe small-sized objects, a microscope equipped with lenses is used in order to realize a magnification of the object when it is viewed through the objective of the apparatus. The object must be illuminated, which is done either by reflection or by transmission, the light reflected or scattered by the object being then recovered by the objective, which forms an image that can be observed. It is in 1873 that E. Abbe [ABB 73] formulated his mathematical theory concerning images formed by a microscope and showed that a point could not be imaged by a point. While working on the theory, he identified a limit to the size of the observable object, which was due to the resolution d = 0.61λ/ NA, where the numerical aperture is NA = 2nsinα. It is expressed as a function of the maximal angle α between the optical axis and the rays coming from the object viewed through the objective under which various rays lead to the formation of the image of the object; n is the refractive index of the environment in which the object is immersed. This resolution limit is nowadays known as the Rayleigh criterion and is associated with the wavelength λ of the radiation used for the observation. It is due to interference effects that lead to the formation of the Airy disk, which for an aperture of diameter D is given by 1.22λ/D under incoherent illumination. The same phenomenon accounts for the "waist" present in the Fabry–Pérot setup used in a laser device, as described in Chapter 6 of [DAH 16] (Figure 6.2). Resolution is limited to 200 nm in the visible spectrum. Abbe thought however that this limit could be overcome with the forthcoming progress in the understanding of physical phenomena or technology.

Indeed, around 55 years later, in 1928, E.H. Synge [SYN 28] proposed a thought experiment that laid the bases of near-field microscopy of modern optics, considering that it was possible to obtain resolutions beyond the limits imposed by diffraction. He considered that it was possible to observe by transmission the interaction of light and matter between an object and the light that would reach the object through a small aperture of dimensions smaller than the wavelength, of the order of 50 nm to 100 nm, in a screen that scans the surface of the object, as shown in Figure 1.8. Using this method, the limit of resolution is no longer imposed by the wavelength, but by the aperture through which light passes. In the theory of light diffraction through a small aperture, the components of the far field can propagate, while in the near field, they are evanescent (see the Appendix: Propagation of a Light Ray). In the proximity of the object, spatial frequencies are high. Referring to the works of Fresnel and to Babinet's theorem, a small-size aperture being equivalent to an object of the same dimension from the perspective of diffraction, this principle prefigures the use of a tip to probe the evanescent waves near a surface.

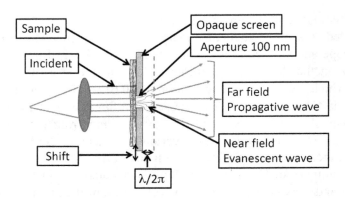

Figure 1.8. *E.H. Synge experimental setup. For a color version of this figure, see www.iste.co.uk/dahoo/metrology1.zip*

Similar to nanotechnologies, it took a while for the idea to be put into practice. In this case, four decades were needed to obtain a better resolution with a scanning microscope developed in 1972 [ASH 72], in the domain of microwaves with a wavelength of 3 cm. A grating could be resolved with a resolution of $\lambda/60$ by this technique.

But meanwhile, scanning tunneling microscope (STM) was invented by Binnig and Roher in 1981 [BIN 82a, BIN 82b, BIN 83, BIN 86a]. The operating principle of STM, which uses the scanning method, relies on the possibility to establish a tunneling current between a metal tip and a metal or conductive surface when a potential difference is maintained between the tip and the surface. The intensity of this current depends on the distance between the tip and the surface. When the tip scans the surface, there is a variation of the distance between the surface and the tip due to surface topology, and consequently the tunnel current, which depends on this distance, also varies. In this case, the current variation is a measure of the surface topology. A second operating mode involves maintaining a constant distance between the surface and the tip. Taking a starting point as a reference, a rated current is fixed. Given that the current variation is positive when the distance between the tip and the surface decreases, and that it is negative when the distance increases, this variation is injected into a feedback system (negative feedback loop) in order to maintain a distance that is equal to the reference distance. The positive or negative correction of the position, made by moving the tip using a piezoelectric system [SYN 32] that expands or contracts when subjected to a potential difference (see Chapter 7 on intelligent materials), is in this case directly related to the topography of the surface. It is worth noting that nowadays STM belongs to the class known as scanning probe microscopy (SPM), but its use is limited to electrically conductive materials.

Finally, in 1984, the first near-field optical microscope is developed [POH 84] in the visible spectrum. The operating principle of scanning near-field optical microscope (SNOM) or near-field optical scanning microscope (NSOM) relies on the Synge principle, but uses an STM, with a probe instead of an opaque screen with an aperture, that is localized in the near-field zone. At nanometric distances, the movement of the probe on a surface is in fact easier to control than that of an opaque plate with an aperture.

A brief description of a SNOM by transmission is provided, as schematically represented in Figures 1.9. and 1.10, highlighting the main elements required for its operation. In a typical optical setup, a laser beam, after passing through neutral filters, is collimated, so that the light rays are quasi-parallel, which corresponds to the propagation of a plane wave (Figure 6.2, Chapter 6 of [DAH 16]). It is then directed under the acceptance angle in an optical fiber whose tip end is near the surface to be analyzed. An optical system is used to recover the signals collected in near field on an APD (Avalanche Photo diode) detector for an analysis through a spectrometer and a processing software. The main elements contributing to data collection are the tip-shaped probe and the precise measurement of the probe movement with respect to the sample by means of a piezoelectric plate subjected to a potential difference. The latter which drives the tip along X, Y or Z is a function of the error signal generated by the laser detection system with a PSD (Position Sensitive Detector) quadrant, which gives the tip position with respect to the surface during the lateral scanning generated by the piezo stage. This scanning is in raster mode, similar to the scanning mode on a television screen. The feedback loop system operates in two modes: at constant force or at shearing force.

In Figure 1.9, the setup of the piezo stage corresponds to the constant force mode. In this case, scanning can be conducted in 3 modes. The contact mode corresponds to the situation in which the tip touches the surface of the sample and maps the topography of the surface. In this mode, there is a risk of sample damage. The tapping mode corresponds to a flexible tip that undergoes deflection according to the topography of the surface. The laser-PSD system measures the deflection angle and generates an error signal, which is used to activate the piezo stage by the feedback loop in order to bring the surface to the reference nominal distance. Finally, the non-contact mode corresponds to that of a high-frequency generator being associated with the tip under oscillation. The amplitudes of these oscillations are modified according to the topography of the surface, and depending on whether the surface gets closer or farther, the vibrations are damped or relaxed by long-range (dispersion) forces between the surface and the tip. Depending on the decrease or increase in vibration amplitudes, the feedback system modifies the distance between the tip and the surface.

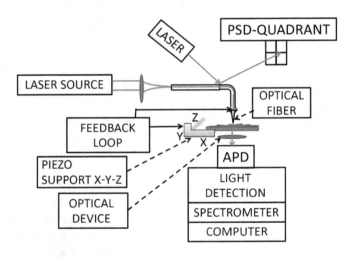

Figure 1.9. *Diagram of a SNOM by transmission in constant force mode. For a color version of this figure, see www.iste.co.uk/dahoo/metrology1.zip*

Another possible setup is that of the tip-carrying fiber being attached to a tuning fork, which sets the tip in a lateral vibration movement above the surface (Figure 1.10). In this shearing force mode, the positioning principle is similar to the above-described contactless mode. Hence, similar to an STM that gives the possibility to probe the metallic surfaces at nanometer scale, the near-field optical microscopy can be used to reach sub-wavelength resolution. A local probe detects a non-radiative signal confined in the close vicinity of the surface, similar to the tunnel current, which is active on the sample surface. Since the probe scans the sample surface at a distance or height of several nanometers, similar to the STM, a negative feedback loop is used to follow the topography of the surface.

Nanometer Scale 25

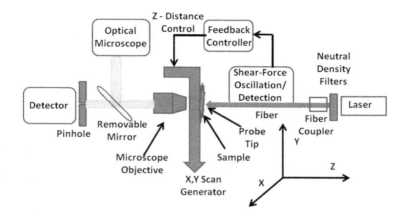

Figure 1.10. *Diagram of transmission SNOM in shear mode. For a color version of this figure, see www.iste.co.uk/dahoo/metrology1.zip*

The atomic force microscope (AFM) was invented in 1986 by Binning, Quate and Gerber [BIN 86b] in a collaboration between IBM and Stanford University. A more detailed discussion on the use of this instrument is provided in the study of thin films of magnetic oxides prepared by laser ablation in section 1.3.2.

In summary, the types of microscopes commonly used nowadays correspond to tunneling and atomic force near-field microscopes, referred to as STM and AFM, respectively. The very high spatial resolution characterizing near-field techniques is related to the very small size of the probe – in the two cases considered, the nanometric apex of a conductive or semi-conductive metallic tip brought to near-contact (a fraction of nanometer) with the surface of the sample. These microscopes use the localized interactions of the tip with the sample and their rapid variation of intensity with the probe–sample distance: the passage of the tunneling current for the STM from which a mapping of the surface atoms is obtained through their local electronic density, and the forces acting on the tip (e.g. Van der Waals interactions for the AFM).

These near-field microscopies give also access to further information, such as electronic spectroscopy by STM and mapping of force fields for AFM and its homologues. The downside of the spatial resolution of these microscopies is the limitation of the analysis to the surface of the sample and to the zones of limited dimension, STM use being also limited to the study of conductive samples. Moreover, the probe may have an invasive effect.

It is worth noting that optical microscopy in the context of light–matter interaction can also be implemented in far field, being in fact less invasive and making it possible to observe (imaging, localization, detection) the sample. Moreover, many spectroscopic signatures can be studied through various spectroscopies using a classical source or a laser as source (Chapter 6, [DAH 16]) such as Brillouin or Rayleigh scattering, absorption or emission spectroscopy (fluorescence or phosphorescence) or Raman spectroscopy, the pump-probe spectroscopy, the double resonance used for probing the properties of matter through the various sub-structures of its energy levels. This spectroscopic analysis can be further extended to the acquisition of physical and chemical parameters such as the lifetime of electronic states. The reader who is interested in the developments of these techniques can refer to the list of references [LEW 833, BET 86, BET 87, FIS 88, COU 89, RED 89].

1.3.1. *Scanning electron microscope*

Electron microscopes are based on the wave–particle duality of the electron Reducing the wavelength of the radiation provides a better resolution than conventional optical microscope. Historically, the first revolution in the field of very high-resolution imaging is that of scanning and transmission electron microscopies (SEM and TEM) (RUSKA 1931), which provide images with nanometer and even atomic resolution due to the nanometer dimension of the electron beam for SEM. Electron microscopy is a very powerful tool, which can also be coupled with spectroscopic chemical analyzes (EELS: electron energy loss spectroscopy) in TEM and structural (diffraction) but their implementation remains very limited, and Electron microscopy is also now widely and mainly used as an imaging tool.

The scanning electron microscope (SEM) relies on the wave properties of an electron to obtain magnification. Figure 1.11 shows the schematic representation of an electron microscope in TEM and SEM modes.

Nanometer Scale 27

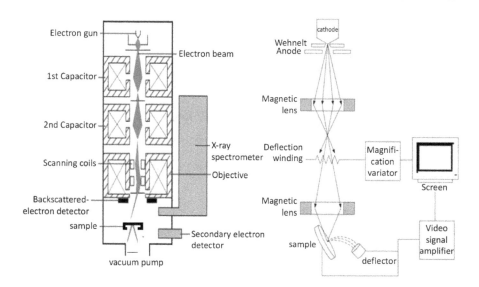

Figure 1.11. *Electron microscope: TEM and MEB modes. For a color version of this figure, see www.iste.co.uk/dahoo/metrology1.zip*

A beam of mono-kinetic electrons is produced by a tungsten filament through which a current is generated by the thermionic effect. Electrons are accelerated by a potential difference V of the order of 100 keV. The wavelength λ associated with electrons is given by the de Broglie relation, $\lambda = h/(\sqrt{2m_e eV})$, where m_e designates the mass of the electron and e its charge. Under 100 keV, the electron wavelength λ is 0.0037 nm. This beam is directed and focused by magnetic lenses to obtain a very small-diameter probe (about 0.01 µm) that scans the surface of the studied sample. Because electrons penetrate the material, the resolution is however not good (rather of the order of a micron for backscattered electrons and of the order of several tenths of a micron for secondary electrons). Electron scattering through the studied sample can be either elastic (no energy loss upon interaction with the nucleus and electrons inside the atoms), which leads to diffraction or plastic (effect of heterogeneities such as grain material boundaries, dislocations, defects, density variations, etc.), which leads to a variation in the intensity of the transmitted beam. The two modes of operation of a TEM are the image mode and the diffraction mode.

The interaction between the beam of electrons and the material generates various types of radiation:

– backscattered electrons;

– secondary electrons;

– absorption generating a current in the sample;

– Auger electrons related to atom de-excitation;

– X-rays emitted during de-excitation following the ionization of electron shells inside the atoms.

The studies of thin films of magnetic oxides were conducted with a Philips CM20 TEM and a Jeol 200CX TEM operating at 200 kV. The two instruments were equipped with a sample support that could be tilted along two axes. Adjustments were directly made on the image displayed by the screen of the image intensifier. Sample preparation involves thin-film manufacturing by ion bombardment perpendicular to the surface (section of the plates). Due to the ion tendency to melt the material, a proper control of this stage is required to avoid sample damage. For the analysis of sections, the surfaces of two pieces of epitaxial layers are glued with epoxy. The section is then mechanically polished on both sides until reaching a thickness of 30 µm and then carefully thinned down by a flow of low-speed argon ions on the samples (6 kV and 10 mA) on a liquid nitrogen-cooled support. This process prevents damages. A surface of the order of 20 µm^2 can then be analyzed by TEM [DAH 04a]. Figures 1.12(a) and (b) and 1.13(a) and (b) show images obtained by TEM for SmFeO$_3$ materials deposited at 785°C and 820°C, respectively.

Figure 1.12. *Diffraction images obtained by TEM for SmFeO3 samples deposited at: a) 785°C and b) 820°C*

TEM-based diffraction analysis reveals two types of structures depending on the temperature deposition of the SmFeO$_3$ film on a quartz substrate: a cubic structure (Figure 1.13(a)) and an orthorhombic structure (Figure 1.13(b)). The deposition has a columnar form in volume with surface extrusions, as shown in Figure 1.12(b) and Figure 1.13(a) and (b).

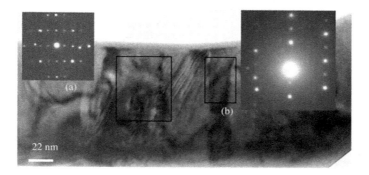

Figure 1.13. *TEM diffraction images for SmFeO3 films deposited at: a) 785°C and b) 820°C*

Figure 1.14(a) and (b) displays SEM images of SmFeO$_3$ films deposited on a quartz substrate at a temperature of 915 C. It is worth noting the presence of protuberances whose dimensions range between 150 and 500 nm, as observed in TEM, and cracks due to different expansion coefficients of the substrate and of the oxide layer, ranging between 50 and 90 nm.

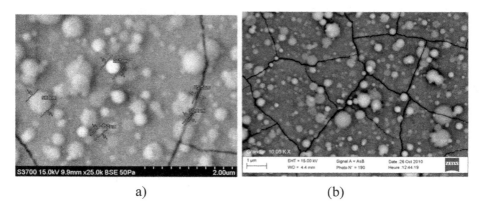

a)   (b)

Figure 1.14. *SmFeO film deposited at TD = 915°C: a) protuberances between 150 and 500 nm; b) cracks between 50 and 90 nm*

1.3.2. *Atomic force microscope*

The atomic force microscope (AFM) is used to obtain three-dimensional analyses of layer roughness with a good vertical resolution. Its operating principle is represented in Figure 1.15 and involves the measurement of the variations of forces of interaction between a tip and the surface under study.

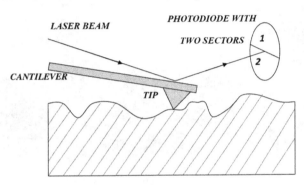

Figure 1.15. *AFM operating principle*

An atomic force microscope is constituted of a piezoelectric quartz plate on which the sample to be studied is placed. The quartz plate lies on an "elevator" engine, and above the sample, there is a very hard tip (read head) fixed at the end of a beam in a frame. A laser beam is reflected on the tip and illuminates two photodiodes. The tip tilt on the scanned surface is determined by the difference in light intensity on the photodiodes. The variation of this electrical signal when the plate under the tip is under translational motion is processed by a software that consequently deduces the topography of the scanned surface.

Figure 1.16 shows the various constituents of an AFM microscope:

– The tip: made from silicon nitride, it is the finest possible, with a pyramid shape.

– The engine: its main role is to lift the sample towards the tip.

– The quartz: it is first of all used for the precise adjustment of the tip on the deposition. Due to the voltage across it (due to the crushing of the tip), it can be used for the computer representation of the deposition surface. Two other voltages are applied to it in order to move the sample in the two directions of the plane.

– The laser: it has a low power – its maximal power is 5 mW. Its essential characteristic is the beam fineness. It emits light in the red band with a wavelength of 670 nm.

– The photodiodes: they generate an electrical signal that is proportional to their respective illumination and thus serve as "sensor of approach" of the tip to the deposit.

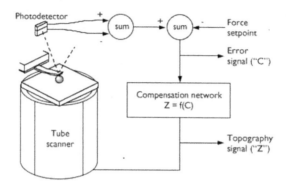

Figure 1.16. *Atomic force microscope*

The work is conducted in the contact mode, the forces being measured in the short range "repulsive" mode.

The laser beam is directed to a metallized part of the cantilever that scans the sample surface horizontally. Due to the deflection of the cantilever, the forces between the tip and the surface can be obtained by the deviation of a laser beam reflected on the metallized part of the cantilever and measured by a quadrant diode (1 and 2). The position of the sample is adjusted along the vertical axis depending on the signal on the diode. Prior to starting a measurement, the laser beam illuminates each quadrant with the same intensity (Itot = I1 - I2 = 0). A vertical variation of the tip position leads to a shift of the beam on the detector and the appearance of the measurement signal (Itot ≠ 0). The signal Itot is then used to vary the position of the piezo-electric engine in order to bring the cantilever to the initial position (Itot = 0). The recorded scanning traces are processed and transformed by a special software into three-dimensional images representing the studied surface.

There are several artifacts for surface measurement with AFM. The most inconvenient in the case of thin films are real geometry and the radius of curvature of the tip, as well as the inertia of the feedback of the piezo-electric engine in case of very high roughness. Since a measurement is the result of a convolution between the geometry of the tip and the irregularities of the surface, the shape of the grains observed on an AFM image is not always actual. To avoid the effect of this artifact, scanning under several angles is recommended. The lateral resolution of an AFM image depends on several factors, among which is the morphology of the surface (notably the size of the grains, their amount and the distance between neighboring grains) and the radius of curvature r_c of a tip. For the tip with $r_c = 50$ nm, the lateral resolution for two grains of height 20 nm is 38 nm. This means that, in order to optically resolve these two grains, the distance between them should be greater than or equal to 38 nm.

The algorithm analyzing the roughness of the film surface is the following: first, a scanning of the maximal possible surface (10×10 μm) is conducted, then, if needed, a zone without large grains or agglomerates (scanning surface typically 800×800 nm) is chosen and roughness is studied. After having repeated this process several times on different areas of a layer, the mean value of the roughness and the mean size of the grains for polycrystalline layers are obtained.

Images of 2×2 μm surfaces on samples deposited under an oxygen pressure of 50 mTorr are displayed in Figure 1.17(a)–(c). Disk- or square-shaped islands of spherical, cylindrical or cubic morphology are observed below the layers deposited at 510°C, 635°C and 755°C. Inspection shows that the size of these islands is smaller at 755°C. The roughness determined on the surfaces of 2×2 μm and 10×10 μm surfaces is in agreement with the roughness obtained by TEM and SE (spectroscopic ellipsometry) [DAH 04a]. The analysis of AFM images of the samples deposited under three different oxygen pressures (24 mTorr, 52 mTorr and 62 mTorr) are displayed in Figure 1.18(a)–(c), for a temperature of deposition of 920°C. Islands of various forms are also observed, their size being smaller when the pressure is lower. Roughness values between 10 and 30 nm are compared to those determined by TEM and SE [DAH 04a].

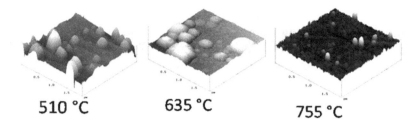

Figure 1.17. *Topology of the deposition surfaces of SmFeO₃ on quartz (P = 50 mTorr) for various deposition temperatures. For a color version of this figure, see www.iste.co.uk/dahoo/metrology1.zip*

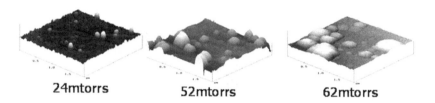

Figure 1.18. *Topology of deposition surfaces of SmFeO₃ on quartz (T = 920°C) for various oxygen pressures. For a color version of this figure, see www.iste.co.uk/dahoo/metrology1.zip*

1.3.3. *Infrared spectroscopy (FTIR/ATR)*

Infrared spectroscopy is used to study the vibrational and rotational motions of molecules in the gaseous phase [DAH 17, DAH 19, DAH 21] and also the vibrational motion of molecules in the condensed phase. Modern characterization instruments rely on the principle of spectroscopic interferometry. It is a non-destructive technique. It can be applied to study the vibration modes of molecules in the medium-infrared-energy domain of polymer materials prepared by spin coating. The identified modes of vibration are used to determine the nature of the chemical bonds susceptible to be present in the molecule. The position (wavelength) and intensity (% in height or surface of the rays) of absorption peaks are specific for various chemical groups. The medium-infrared domain corresponds to the interval $400 \text{ cm}^{-1} < \sigma$ (wave number) $< 4{,}000 \text{ cm}^{-1}$.

The general principle of FTIR spectroscopy is presented in Figure 1.19. The infrared beam coming from the source is directed towards a Michelson interferometer that modulates each wavelength of the beam at a frequency that differs from the frequency associated with IR radiation. In the interferometer, light falls on the lens that splits the beam in two. Half of the beam is directed to a fixed mirror, while the rest is directed to a mobile mirror whose movement modulates the IR beam. When the two beams recombine, destructive or constructive interferences occur, depending on the position of the mobile mirror.

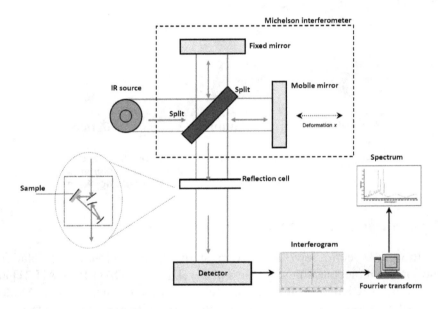

Figure 1.19. *Principle of FTIR spectroscopy, and the cell of measurement in ATR. For a color version of this figure, see www.iste.co.uk/dahoo/metrology1.zip*

The modulated beam is directed towards the sample through a high-index crystal. Absorptions occur in the material in contact with the crystal following an internal reflection and a tunnel effect at the reflection surface. After reflection, the beam returns to a sensor that emits an electrical signal. The latter appears as an interferogram, which is a signal whose amplitude depends on the position of the mobile mirror. This position is determined from a starting position corresponding to an identical position of the two mirrors on the two arms and which corresponds to a zero-path difference between the two beams and the interference fringes obtained with a He–Ne laser beam that travels along the same path. The modulation of the IR signal corresponds to the Fourier transform of the IR spectrum.

ATR (Attenuated Total Reflection) is a technique that is adapted to the infrared spectroscopy study of materials that are too absorbing or too thick to be analyzed by transmission. In ATR spectrometry, the sample to be analyzed, in liquid or solid state, in the form of powder or thin film, is maintained against the basis of a crystal with high refractive index n_1. For powders and solid materials, a small press is used to maintain the contact with the crystal lens. The other types of samples must be flat or elastic in order to best fit the shape of the crystal.

If n_2 is the refractive index of the sample, there is a total reflection of the incident ray if the angle θ between the sample–crystal interface and the crystal facets is greater than the critical angle θ_c given by the relation: $\sin \theta_c = n_2/n_1$. A certain penetration of the infrared beam takes place at the surface of the sample leading to a decrease in the intensity of the reflected beam in the ranges of frequency where the sample features an absorption. The refractive indices of ATR crystals normally range between 2.4 (ZnSe, KRS-5: mixed crystals of thallium bromides and iodides) and 4 (Ge), and the incidence angles of the systems are generally 30°, 45° and 60°. The total optical path depends on the number of internal reflections, which in turn depends on crystal geometry: length, thickness and angle of incidence.

Using a Thermo Scientific Nicolet iS10 FTIR spectrometer with a Thermo Scientific Smart iTR attenuated total reflectance (ATR) sampling accessory, samples are characterized [KHE 14]. The results in Figure 1.20 correspond to the measurements in the spectral region between 500 and 4,000 cm^{-1} with a resolution of 1 cm^{-1} on various solutions studied in [KHE 14] and whose ellipsometry studies are presented in Chapter 8 of [DAH 16].

The same peaks are found on the spectra of all polymers. The valence vibrations of methyl groups are observed for the CH stretching mode in the form of a group of about 2,962 cm^{-1} and 2,904 cm^{-1}; the stretching vibrations are about 1,413 cm^{-1} followed by a small band at 14,40 cm^{-1}. The antisymmetric stretching of siloxane Si–O–Si functions, which have the form of a high peak with a shoulder, is about 1,008 cm^{-1} and 1,082 cm^{-1}. Si–C valence vibrations have the form of a very high peak about 784 cm^{-1} corresponding to the rocking stretching of Si–CH$_3$, a stretching mode at 864 cm^{-1} and a significant band at 1,258 cm^{-1} corresponding to a stretching vibration. The seven samples from different suppliers have the same spectral signature, the only difference being absorption intensities and an additional band for M2. The difference in intensity between the absorption peaks reflects a difference in the concentration of functional groups specific to each supplier in the polymer chain.

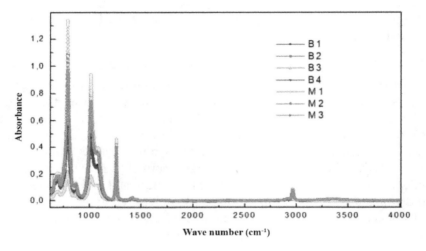

Figure 1.20. *Absorption spectra of silicone gels B (1–4) and M (1–3). For a color version of this figure, see www.iste.co.uk/dahoo/metrology1.zip*

An ATR spectroscopy study of polymers is conducted on samples of polymer subjected to constraints of temperatures and humidity in a super-Highly Accelerated Life Test (HALT) and Highly Accelerated Testing (HAT) device described in Chapter 1 of [POU 20]. Among the characteristics of HALT, which comprises a generator of vibrations and thermal shocks, the following can be listed: a T° slope of 60°/mn, a range of T° between +200° and -100°C and a range of vibration per pistons under the lower plate between 0 and 50 g. The humidity generator can reach a maximal relative humidity of 98%, for a volume of 1,680 liters. To run the aging tests, the test profile software control in terms of temperature, vibration and humidity is activated. The polymer samples were subjected to cycles such as those represented in Figures 1.20 and 1.21.

Nanometer Scale 37

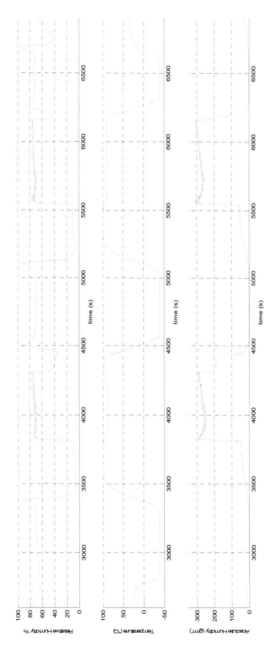

Figure 1.21. *Combined thermal cycling (T: -45 at 95°C) and humidity cycling (HR: 70%) in the super-HAT. For a color version of this figure, see www.iste.co.uk/dahoo/metrology1.zip*

38 Applications and Metrology at Nanometer Scale 1

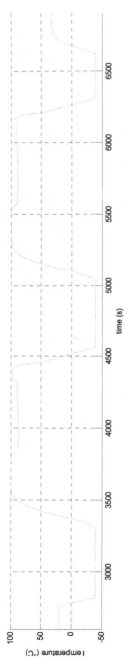

Figure 1.22. *Thermal cycling (T: –45 at 95°C) in HALT. For a color version of this figure, see www.iste.co.uk/dahoo/metrology1.zip*

To study the effect on aging of an environment combining relative humidity with temperature, two types of loads are applied: a (T+H) load (Figure 1.21) combining a temperature cycle (variation of temperature between 45°C and 95°C) with a humid atmosphere (maximum relative humidity rate is 78%) and a (THalt) load consisting of thermal cycles (Figure 1.22).

Another type of aging due to a more classical aggressive thermal stress is conducted on polymers in a furnace. The various stress temperatures are 150°C, 180°C and 200°C. The monitored thermal profile is programmed by a progressive increase in temperature from 25°C to the temperature to be reached during 15 minutes followed by a plateau of 2 h at this temperature. All the samples are subjected to the same process under the same conditions. Figure 1.23 displays the applied thermal cycles.

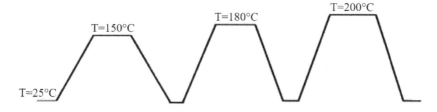

Figure 1.23. *Thermal cycling in a furnace*

The spectra were obtained by ATR in the interval between 600 cm^{-1} and 4,000 cm^{-1}. The analyses show a differentiation in the results in two different spectral regions located between 600 cm^{-1} and 1,500 cm^{-1} and between 2,280 cm^{-1} and 2,400 cm^{-1}. The first interval corresponds to the region of Si–O vibration modes and the second interval to the spectral signature of CO_2 vibration modes, indicating the presence of this molecule that is trapped in nanocages in the polymer structure.

Figure 1.24 shows the spectra obtained with M-type polymers (1 component). Figure 1.25 shows the spectra obtained with B-type polymers (2 components) in the region of Si–O, between 600 cm^{-1} and 1,500 cm^{-1}. The spectra are denoted by {TF} for the thermal stress applied in a furnace, {Thalt} for the thermal stress in a dry environment and {T+H} for the thermal stress in a humid environment.

40 Applications and Metrology at Nanometer Scale 1

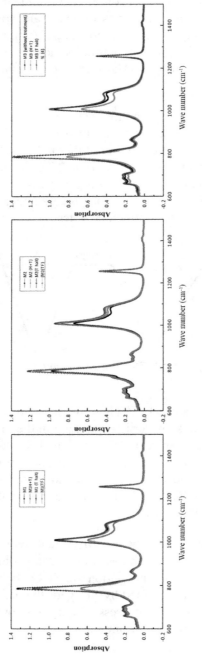

Figure 1.24. *ATR spectra of M-type polymers. For a color version of this figure, see www.iste.co.uk/dahoo/metrology1.zip*

Nanometer Scale 41

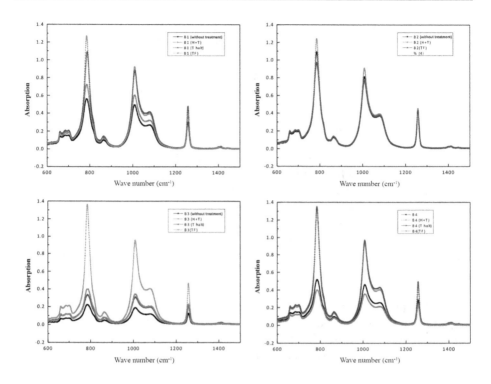

Figure 1.25. *ATR spectra of B-type polymers. For a color version of this figure, see www.iste.co.uk/dahoo/metrology1.zip*

The intensities of absorption peaks vary for each material, but in different proportions. The absorption increases or decreases depending on the material and the TF, THalt or {T+H} loads. As a general rule, the increase or decrease in the intensity of an absorption peak by thermal processing influences also all the peaks of the spectrum in the same direction of variation. Either an increase or a decrease in the intensity can be observed throughout the spectrum, but observing a decreasing peak and an increasing peak and vice versa is excluded. No general tendency of evolution of the intensities of absorptions depending on the thermal processing can be globally established.

The effect on the spectral signatures of {THalt} and {T+H} loads is more or less similar in the two groups of M-type or B-type polymers. In the B-types, the peaks are rather more intense or of the same order as for the M-type. Comparing the effect of the three thermal processes, the spectral signature of the TF effect corresponds to signals of lower intensity than for {THalt} and {T+H,} effects, with the exception of the two compounds M2 and B2 with equivalent signatures.

Concerning the zone located in the region of intense absorption of the antisymmetric stretching vibration of CO_2, between 2,280 cm^{-1} and 2,400 cm^{-1}, the absorption spectra have the same evolutions in the intensities of the peaks of absorptions before and after processing, independently of each material. This analysis yields different results from those obtained on the spectra from 600 cm^{-1} to 1,500 cm^{-1}. Relying on the intensity of absorption bands centered at 2,350 cm^{-1}, the following characteristics are listed.

All the polymers feature a quasi-null absorption in the IR spectrum at a very low level (Absorbance $\approx$ 0) before the application of the thermal constraint. On the other hand, the thermal load in a humid environment {T+H} leads to an increase in absorbance, contrary to the {THalt} and {TF} loads that lead to a decrease in absorbance (Absorbance < 0).

A study of possible absorptions in this region reveals the absorption by triple bonds (C≡O, C≡N) from 2,000 cm^{-1} to 2,500 cm^{-1}, by double bonds (C=O, C=N and C=C) from 2,000 cm^{-1} to 1,500 cm^{-1} and more specifically, the case of a double bond (O=C=O) as in CO_2. For the latter, it is worth noting the presence of a vibration–rotation band (Branch P and R) centered in the gaseous phase at 2,349.14 cm^{-1} due to the absorption of the mode of vibration v_3 of antisymmetric stretch of CO_2. This corresponds with the identified zone taking into account the effect of nanocage perturbation [DAH 99, DAH 06] on the motion of the molecule as a result of the interaction with the electrical field present in the polymer. In order to evidence the presence of CO_2 in the polymers, the difference spectra are drawn using the spectrum recorded after applying the most aggressive thermal load {TF} as a basic line. The curves are represented in Figures 1.26 and 1.27 for type M and B polymers, respectively.

Nanometer Scale 43

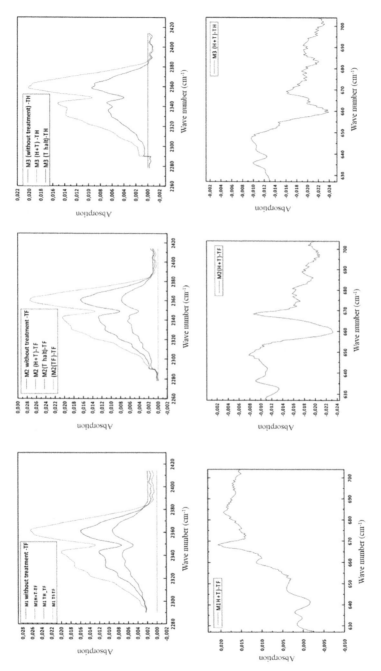

Figure 1.26. *ATR difference spectra for M-type polymers. For a color version of this figure, see www.iste.co.uk/dahoo/metrology1.zip*

44 Applications and Metrology at Nanometer Scale 1

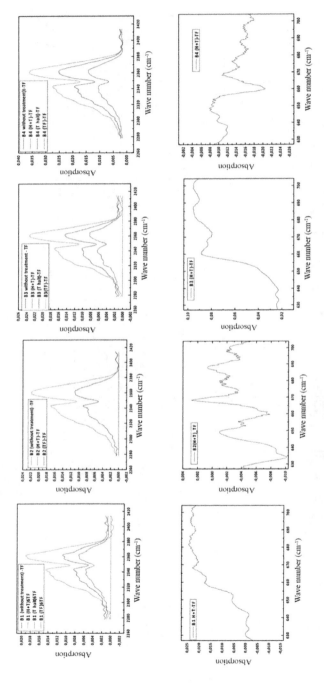

Figure 1.27. *ATR difference spectra for B-type polymers. For a color version of this figure, see www.iste.co.uk/dahoo/metrology1.zip*

The difference spectra clearly show the absorptions of CO_2 in the region of absorption in the band associated with the antisymmetric stretching approximately 2,350 cm^{-1} characterized by a branch P (low frequency) and a branch R (high frequency). The gap between the maximum of each branch is about 22 cm^{-1}. This gap is given by the formula $\Delta v = 2,358\sqrt{TB}$ [HER 45], where Δv is expressed in cm^{-1}, B is the rotational constant of CO_2 and T is the absolute temperature. Considering a value of 0.39 cm^{-1} for B and 293 K for T, a gap of 25 cm^{-1} is calculated, which is comparable to the value measured on the spectra. In the region of atmospheric absorption of the bending mode band v_2, the peak at 668 cm^{-1} can be observed, particularly on M-type polymers. This value can be compared with that of 667 cm^{-1}, which corresponds to the absorption in gaseous phase of this vibration that is doubly degenerate. Certain spectra feature a doublet at 650 cm^{-1}, 655 cm^{-1}, which might be associated with a degeneracy removal.

CO_2 could be simultaneously considered a Lewis acid (positive charge on the central carbon, electron acceptor) and a Lewis base (negative charge on the terminal oxygen atoms, electron donors). According to the works conducted on CO_2 in polymers [NAL 05, NAL 06, CUL 13], degeneracy removal is due to an interaction of the type Lewis acid–base between CO_2 and the groups in the polymer (ester, ether, aromatic ring, carbonyl group, etc.). Degeneracy removal is due to interactions with the carbonyl group, aromatic rings or sites with basic properties.

1.4. Conclusion

This chapter presents various deposition techniques that are used for material nano-structuring, as well as the equipment developed for sample characterization and analysis. In situ during the elaboration phase, these techniques are used to control the fabrication of materials on the nanometer scale. Some of these techniques can be used to manipulate matter at the scale of atoms. A detailed study using ATR spectroscopy, through which a material is probed by the tunnel effect over a nanometer-scale thickness, is presented. The material being pressed firmly into contact on a high-index crystal, the beam returning after total reflection is used to obtain an IR beam containing information on the spectral signature of the constituents trapped in the polymer. Absorption of CO_2 is revealed in the polymers used as encapsulation material.

1.5. Appendix: light ray propagation

In light diffraction problems, the amplitude of a wave $E_z(x,y)$ on a surface located in a plane z is determined from the amplitude $E_0(x,y)$ of a wave in the plane $z = 0$.

In Chapter 3 of [DAH 16], equation 3.6 gives the expression of an electromagnetic wave moving along Oz, in the form u (z-vt) = acos (k(z-ct)), where v is the wave propagation speed in a medium of index n, a its amplitude and k = 2 πn/λ. In a vacuum, n = 1 and in a three-dimensional space, the monochromatic plane wave of angular frequency ω = kc and of wave vector **k** = (kx, ky, kz) can be expressed in complex notation, in the form **E** = **E₀** exp i(2π/λ (αx+βy+γz) - ωt), where the components of the wave vector are in the form $k_x = 2\pi\alpha/\lambda$, $k_y = 2\pi\beta/\lambda$ and $k_z = 2\pi\gamma/\lambda$, with α, β and γ being the direction cosines of the wave vector **k** and λ the wavelength. This expression can be obtained by solving the Helmholtz equation (equation 3.7, [DAH 16]):

$$\Delta \vec{E} + \frac{\omega^2}{c^2} \vec{E} = \vec{0} \qquad [1.1]$$

where $\Delta \vec{E} = \vec{\nabla}^2 \vec{E}$, using the Green function $G_k(\vec{r}, \vec{r}_0)$ of the Helmholtz equation, which verifies that:

$$(\nabla^2 + k^2)G_k(\vec{r}, \vec{r}_0) = \delta(\vec{r} - \vec{r}_0) \qquad [1.2]$$

given that: $G_k(\vec{r}, \vec{r}_0) = -\frac{exp(ik|\vec{r}-\vec{r}_0|)}{4\pi|\vec{r}-\vec{r}_0|}$, and the unit vector $\vec{e}_r = \frac{(\vec{r}-\vec{r}_0)}{|\vec{r}-\vec{r}_0|}$. The solution is straightforward, and for the k mode of the electric field from a source placed in $\vec{r}_0$, we can write: $\vec{E}_k(\vec{r}, \vec{r}_0) = -\frac{exp(ik|\vec{r}-\vec{r}_0|)}{4\pi|\vec{r}-\vec{r}_0|} \vec{e}_r$.

The propagation of a wave through objects (transparent apertures or opaque objects) can be determined using the Green–Ostrogradsky theorem and the Green function $G_k(\vec{r}, \vec{r}_0)$ of the Helmholtz equation.

From the relation:

$$\vec{\nabla}(u\vec{\nabla}v - v\vec{\nabla}u) = u\vec{\nabla}^2 v - v\vec{\nabla}^2 u = u\Delta v - v\Delta u \qquad [1.3]$$

where u and v are solutions of the Helmholtz equation, and from the Green–Ostrogradsky theorem, which transforms a volume integral into a surface integral:

$$\int (u\Delta v - v\Delta u)d\tau = \int (u\vec{\nabla}^2 v - v\nabla^2 u)d\tau = \oint (u\vec{\nabla}v - v\vec{\nabla}u)d\vec{s} \quad [1.4]$$

It can be written that:

$$\int \left(\vec{E}(\vec{r})\Delta G_k(\vec{r},\vec{r}_0) - G_k(\vec{r},\vec{r}_0)\Delta\vec{E}(\vec{r})\right) d\tau = \int \vec{E}(\vec{r})\delta(\vec{r}-\vec{r}_0)d\tau$$

$$- \int \vec{E}(\vec{r})k^2 G_k(\vec{r},\vec{r}_0)d\tau + \int G_k(\vec{r},\vec{r}_0)k^2\vec{E}(\vec{r})d\tau = \vec{E}(\vec{r}_0) \quad [1.5]$$

Hence

$$\vec{E}(\vec{r}_0) = \oint \left(\vec{E}(\vec{r})\vec{\nabla}G_k(\vec{r},\vec{r}_0) - G_k(\vec{r},\vec{r}_0)\vec{\nabla}\vec{E}(\vec{r})\right) d\vec{s} =$$
$$\oint (\vec{E}(\vec{r})\{\vec{\nabla}G_k(\vec{r},\vec{r}_0)d\vec{s} - ikG_k(\vec{r},\vec{r}_0)ds\}) \quad [1.6]$$

where the surface integral bounds the point located at $\vec{r}_0$ and $\vec{r}$ is the position vector of the surface element $d\vec{s}$ (Figure 1.28), such that $\vec{\nabla}\vec{E}(\vec{r}).d\vec{s} = ik\vec{E}(\vec{r})ds$, for a wave that propagates inwards in the volume bounded by the integral surface element ds. For a wave propagating outwards of the surface, $\vec{E}(\vec{r}).d\vec{s} = -ik\vec{E}(\vec{r})ds$.

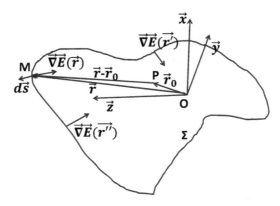

Figure 1.28. *Kirchhoff integral on a surface Σ passing through a point M enclosing the point P. For a color version of this figure, see www.iste.co.uk/dahoo/metrology1.zip*

Given that $G_k(\vec{r}, \vec{r}_0) = -\frac{exp(ik|\vec{r}-\vec{r}_0|)}{4\pi|\vec{r}-\vec{r}_0|}$, it can be written that:

$$\vec{\nabla}G_k(\vec{r}, \vec{r}_0) = ikG(\vec{r}, \vec{r}_0)\vec{e}_r + \frac{exp(ik|\vec{r}-\vec{r}_0|)}{4\pi|\vec{r}-\vec{r}_0|^2}\vec{e}_r = \left(\frac{ik}{|\vec{r}-\vec{r}_0|} - \frac{1}{|\vec{r}-\vec{r}_0|^2}\right)(\vec{r} - \vec{r}_0)G_k(\vec{r}, \vec{r}_0) \qquad [1.7]$$

At a great distance compared to the wavelength such that $k|\vec{r} - \vec{r}_0| \gg 1$, the following solution is obtained:

$$\vec{E}(\vec{r}_0) = \oint \left(\vec{E}(\vec{r}) \left\{\frac{ik}{|\vec{r}-\vec{r}_0|}(\vec{r} - \vec{r}_0)G_k(\vec{r}, \vec{r}_0)d\vec{s} - ikG_k(\vec{r}, \vec{r}_0)ds\right\}\right) \qquad [1.8]$$

such that:

$$\vec{E}(\vec{r}_0) = -\frac{ik}{4\pi}\oint \vec{E}(\vec{r})\frac{exp(ik|\vec{r}-\vec{r}_0|)}{|\vec{r}-\vec{r}_0|^2}\left((\vec{r} - \vec{r}_0)d\vec{s} - |\vec{r} - \vec{r}_0|ds\right) \qquad [1.9]$$

This solution is the mathematical formulation of the Huygens–Fresnel principle or the Huygens–Fresnel equation obtained by the Kirchhoff integral. Hence, it is the Fresnel–Kirchhoff diffraction formula.

When a monochromatic plane wave propagating parallel to Oz encounters a screen (E_0) with an opening (Figure 1.29), a diffraction pattern is formed in the observation plane. To simplify, the plane of the opening is supposed to be perpendicular to the propagation of the plane wave and that the amplitude and the gradient of the electric field of the wave are constant on the surface of the opening located in the vicinity of M (Figure 1.28) and zero elsewhere. The amplitude on the other side of the aperture at a point P located at $\vec{r}_0$ is given by the Fresnel–Kirchhoff diffraction formula. If the opening is in the Oxy plane (Figure 1.29), and the dimensions are small compared to the position of the point P located at $\vec{r}_0$, the amplitude of the field at P writes as:

$$\vec{E}(\vec{r}_0) = -\frac{ik\vec{E}(\vec{r})}{4\pi R}(cos\theta + 1)\oint exp(ik|\vec{r} - \vec{r}_0|)dxdy \qquad [1.10]$$

where R is the distance between the center of the opening and the point P (as a first approximation, it is the same distance for all the points of the opening), and θ is the angle between the axis Oz and the vector $\overrightarrow{MP} = \vec{r}_0 - \vec{r}$. The origin of the coordinates system is the plane containing the diffracting opening.

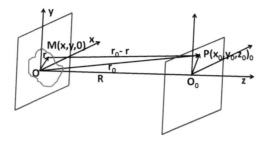

Figure 1.29. *Schematic of the diffraction by an aperture scanned by M in a plane at point P. For a color version of this figure, see www.iste.co.uk/dahoo/metrology1.zip*

The modulus of vector $\vec{r}_0 - \vec{r}$ is given by: $|\vec{r}_0 - \vec{r}| = \sqrt{r_0^2 - 2\vec{r}_0 \cdot \vec{r} + r^2}$.

Under the far-field Fraunhofer diffraction conditions, with an origin O in the plane of the aperture (Figure 1.29), where R is the distance between the observation plane and the diffracting plane, we can write that:

$$exp(ik|\vec{r} - \vec{r}_0|) = exp\left(ik\sqrt{r_0^2 - 2\vec{r}_0 \cdot \vec{r} + r^2}\right)$$
$$= e^{ikR} exp(-ik(\alpha x + \beta y)) \qquad [1.11]$$

where: $\alpha = \frac{x_0}{R}$ and $\beta = \frac{y_0}{R}$.

In the case of the Gaussian approximation, with paraxial light rays, the field at P is expressed as:

$$\vec{E}(\vec{r}_0) = A \oiint exp(-ik(\alpha x + \beta y))dxdy \qquad [1.12]$$

We can thus calculate the Fraunhofer diffraction at a far distance by this formula. In the case of a circular aperture, it can be written as:

$$\vec{E}(\vec{r}_0) = A \oiint exp(-ik(\alpha\rho\cos\varphi))\rho d\rho d\varphi$$
$$= 2\pi A \int_0^a J_0(k\alpha\rho)\rho d\rho = 2\pi A a^2 \frac{J_1(k\alpha\rho)}{k\alpha\rho} \qquad [1.13]$$

where $J_0(x)$ and $J_1(x)$ are the Bessel functions of zeroth and first orders, respectively.

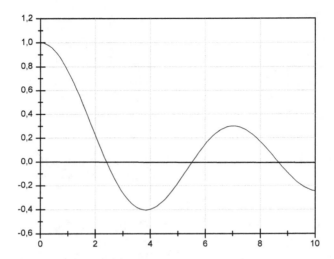

Figure 1.30. *Diffraction amplitude distribution by a circular aperture. For a color version of this figure, see www.iste.co.uk/dahoo/metrology1.zip*

Fresnel diffraction is localized in an observation plane closer to the aperture than a distance of $\lambda/2\pi$. In this case, the terms of order 2 in x and y must be considered and it is wise to use a coordinate system so as to get rid of the terms of order 1 in x and y. The aperture is placed in the Oxy plane with the Oz axis, which is perpendicular to the plane passing through the observation point r_0. In this coordinate system, $\alpha = \beta = \vec{r}.\vec{r}_0 = 0$. Under such conditions, the following expression is calculated:

$$exp(ik|\vec{r} - \vec{r}_0|) = exp\left(ik\sqrt{r_0^2 - 2\vec{r}_0.\vec{r} + r^2}\right) = e^{ikR} exp\left(\frac{ik(x^2+y^2)}{2R}\right) \quad [1.14]$$

In the case of the Gaussian approximation, with paraxial light rays, the field at P is expressed as:

$$\vec{E}(\vec{r}_0) = A \int exp\left(\frac{ikx^2}{2R}\right) dx \int exp\left(\frac{iky^2}{2R}\right) dy \quad [1.15]$$

From which it is possible to calculate the Fresnel diffraction pattern.

2

Statistical Tools to Reduce the Effect of Design Uncertainties

The inclusion of the ability to resist potential design or process variations is an innovative product design approach. This includes managing critical parameters of the system, using security coefficients and applying more advanced techniques for reliability optimization. This approach consists of applying statistics to design a system that has the best performance although its parameters have potential variations. The designed product respects the required performance in spite of uncertainties and remains above a threshold of minimal performance for a given probability. These innovative engineering design methods are related to multi-objective optimization problems. This chapter presents several methods and practical applications for improving the system design by including potential uncertainties.

2.1. Introduction

When a product is mass manufactured, it is obvious that design parameters must respect both manufacturing tolerances and expected variable final use conditions. The corresponding unknown parameters must be taken into account during product design. The statistical tools that include system uncertainties rely on:

– developing an approximate mathematical model of the physical system under study;

– identifying and characterizing the sources of uncertainty among the parameters of the model;

– studying the propagation of uncertainties and their impact on the output signal (response) of the system.

The methods for propagation of uncertainties depend on the mathematical tools that are used. The principal methods applied are the reliability-based design optimization, the probabilistic approach based on design of experiments and the set-based approach. A post-processing stage offers the possibility of analyzing and statistically characterizing the response (statistical moments, distribution, etc.).

2.2. Review of fundamental definitions in probability theory

2.2.1. *Definitions and properties*

Fundamental definitions of the probability theory are presented [MOH 10].

Consider a domain D that represents the physical domain on which the problem is defined.

In addition, consider the space H of real-valued functions defined over D:

$$H=\{ f \ /f: D \rightarrow \mathbf{R} \}$$

The inner product over H is defined as:

$$\langle f(x), g(x) \rangle = \int_D f(x).\overline{g(x)}dx \qquad [2.1]$$

where $f, g \in$ H and g^* is the complex conjugate of g.

With this scalar product, H is a Hilbert space.

The probability theory seeks to formally model intrinsically random phenomena using random experiments.

A random experiment ε is fully described by the probability triplet (Ω, B, p), which is also referred to as "the probability space associated with the random experiment ε".

Ω: is a random set that represents the set of all possible results of the experiment ε.

B: is the set of events of ε (also known as σ-algebra in the analysis).

p: is the probability defined on (Ω, B).

Therefore, a probability is defined on an arbitrary set Ω associated with a set B of partitions of Ω, which verifies the following properties:

i) $\Omega \in B$.

ii) If $A \in B$, then $\overline{A} \in B$ ($\overline{A}$ is the complementary of "A").

iii) Any finite or countable union of elements of B belongs to B $A_j \in B, \forall j \in N \rightarrow \bigcup_{j=0}^{+\infty} A_j \in B$ and $\bigcap_{j=0}^{+\infty} A_j \in B$.

Hence, the probability of p on (Ω, B) is an application of B in $[0,1]$, which verifies that:

$$\begin{cases} p(\Omega) = 1 \\ \\ p\left(\bigcup_{j \in J} A_j\right) = \sum_{j \in J} p(A_j) \end{cases} \qquad [2.2]$$

for any finite countable set of disjoint events $(A_j, j \in J)$, i.e. for any A_j such that: $A_j \cup A_k = \phi \quad \forall (A_j, A_k) \in \Re^2, j \neq k$.

Consequently, for arbitrary events A and B, we have the following properties:

i) $p(\phi) = 0$

ii) $p(A) = 1 - P(\overline{A})$

iii) $p(A \cup B) = p(A) + P(B) - P(A \cap B)$ $\qquad [2.3]$

iv) $A \subset B \Rightarrow p(A) \leq p(B)$

v) $p(A \cap B) \leq P(A) \leq p(A \cup B)$

Let (Ω, B, p) be a probability space. The space Θ of real-valued measurable functions of Ω is defined as:

$$\Theta = \{g / g : \Omega \to \Re\}$$

The inner product on Θ is defined similarly to the inner product of H:

$$\langle \alpha(\theta) \beta(\theta) \rangle = \int_{\Omega} \alpha(\theta) \beta(\theta) dp, \alpha, \beta \in \Theta \qquad [2.4]$$

2.2.2. Random variables

A real random variable X is a B-measurable application of Ω in $\Re$, which means that for any A $\in B(\Re)$, a Borel set, we have $X^{-1}(A) \in B$. The application X generates a probability P_X on $(\Re, B(\Re))$, which is defined as:

$$P_X(A) = p\left[X^{-1}(A)\right] = p[\omega \in \Omega, X(\omega) \in A] \qquad [2.5]$$

Here "X" is related to a random experiment whose results are real numbers.

Hence, $P_X(A)$ is the probability that the result of the experiment belongs to A.

In the case of discrete random variables, the probability law of X is defined as:

$$P(X = x_j) = P_j, \ j \in J; J \text{ finite countable.}$$

where $P_j > 0$ and $\sum_{j \in J} P_j = 1$; with $x_j \neq x_k$ for $j \neq k$.

$$p(X \in A) = \sum_{j:x_j \in A} P_j$$

For a continuous random variable, the probability law of X is defined by a probability density function $f(x)$ such that:

$$p(X \in A) = \int_A f(x)dx \qquad [2.6]$$

and $\forall x \in \Re \quad f(x) \geq 0 \quad$ and $\quad \int_{-\infty}^{+\infty} f(x)dx = 1 \qquad [2.7]$

2.2.3. Random vectors

A real (or complex) random vector $\{X\}$ of size "n" on the set Ω is defined as a function of Ω for x, the subspace of $\Re^n$ (or C^n):

$$\{X\}(w): \Omega \rightarrow x \subset \Re \,(or\, C)$$

The real random vector $\{X\}$ is a vector of "n" real random variables:

$$\{X\} = \{X_1, X_2 \rightleftarrows, X_n\}^{\,T}$$

The function $\{X\}(w)$ must verify the following conditions:

a) The set $\{w:\{X\}(w) \leq \{x\}\}$ is an event in Ω for $\forall \{X\} \in \Re^n$.

(The notation $\{X\}(w) \leq \{x\}$ means that, for any $i=1,2,\ldots\ldots,n$, we have $X_i(w) \leq x_i$);

b) $p\left(\{w:\{X\}(w) \leq +\infty\}\right) = 1$;

c) $p\left(\{w:\{X\}(w)\le-\infty\}\right)=0.$

The definitions of the probability density function can be extended to vectors of random variables.

Hence, we define the joint probability function of order n, denoted by $f_n\left(x_1,x_2......,x_n\right)$, for a vector of random variables $\{X\}$ of size "n", such that:

$$p\left(a_i\le x_i\le b_i\right)=\int_a^b\int_{a_2}^{b_2}........\int_{a_n}^{b_n}f_n\left(x_1,x_2,...........x_n\right)d_1\,dx_2.......dx_n$$

With the properties:

$$f_{n(x_1,x_2........x_n)}\ge 0$$

$$\underbrace{\int_{-\infty}^{+\infty}\int_{-\infty}^{+\infty}.........\int_{-\infty}^{+\infty}}_{n\ times}f_n\left(x_1,x_2,......x_n\right)=1 \qquad [2.8]$$

where $p\left(a_i\le x_i\le b_i\right)$ is the probability that the random variable X_i ranges between a_i and b_i .

2.2.4. *Static moments*

The moments of one or several random variables are defined as the mathematical expectations of various powers of these random variables. For a single variable, $E\lfloor X^n\rfloor$ is the moment of order "n"; $E\lfloor X^m Y^n\rfloor$ is the joint moment of order "$m + n$" of variables "X" and "Y". Once the probability distribution is known, all the moments can be calculated.

However, without additional information (as is the case for a normal distribution), knowing the moments of all the orders is a prerequisite for obtaining the probability distribution.

If $Y = g(x)$ is a continuous and real function of the random variable X, then Y is also a random variable, and the mathematical expectation of the function $g(X)$ is defined as:

$$E[Y] = E[g(X)] = \int_{-\infty}^{+\infty} g(x)f(x)dx \qquad [2.9]$$

where $f(x)$ is the probability density function of the variable X.

In the specific case of $g(X) = x$, the mathematical expectation $E[g(X)]$ becomes the *expected or mean value* of the random variable X, denoted by $\mu(X)$ or $\bar{x}$, such that:

$$\mu(X) = \int_{-\infty}^{+\infty} xf(x)dx \qquad [2.10]$$

Generally speaking, we can define the K-th static moment of a random variable as:

$$\mu^k(X) = \int_{-\infty}^{+\infty} x^k f(x)dx \qquad [2.11]$$

The centered static moments are defined with respect to the mean:

$$\overline{\mu^k}(X) = E[(X - \mu(X))^k] = \int_{-\infty}^{+\infty} (x - \bar{x})^k f(x)dx \qquad [2.12]$$

The first static moment is the mean defined by equation [2.11]; the second centered moment defines the variance of the random variable.

The variance of the random variable can be defined as:

$$V(X)E(X^2) + (E(X))^2 \qquad [2.13]$$

PROPERTIES.–

$$E(aX+b)=aE(X)+b$$

$$V(aX+b)=b \qquad [2.14]$$

$$E(X-E(X))=E(X)-E(X)=0$$

$$V\left(\frac{X-E(X)}{\sigma(X)}\right)=1 \qquad [2.15]$$

The variance of a reduced centered variable is equal to 1:

$$E\left(\frac{X-E(X)}{\sigma(X)}\right)=0 \qquad [2.16]$$

The expectation of a reduced centered variable is equal to 1.

The positive root of the variance defines the *standard deviation* of the random variable; *the coefficient of variation* of a random variable is given by:

$$\varsigma(X)=\sqrt{\frac{Var(X)}{\mu(X)^2}} \qquad [2.17]$$

The notation of the mathematical expectation also applies to the vectors of random variables. We can therefore define the covariance between two random variables as:

$$Cov(X_i,X_j)=E\left[(X_i-\overline{x}_i)(X_j-\overline{x}_j)\right]$$
$$=\int_{-\infty}^{+\infty}\int_{-\infty}^{+\infty}(x_i-\overline{x}_i)(x_j-\overline{x}_j)dx_i\,dx_j \qquad [2.18]$$

The dimensionless correlation coefficient between two random variables is given by:

$$r_{ij} = \frac{Cov\left(X_i, X_j\right)}{\sqrt{Var\left(X_i\right)Var\left(X_j\right)}}$$

[2.19]

It can be shown that $\left|r_{ij}\right| \le 1$.

The variables 'X_i' and 'X_j' are referred to as uncorrelated if

$r_{ij} = 0$ (or $Cov\left(X_i, X_j\right) = 0$). Consequently, the statically independent random variables are uncorrelated. However, the zero correlation coefficient (null covariance) does not necessarily involve the static independence of the random variables involved.

The extension of these concepts to the random variable vector $\{X\} = \{X_1, X_2, \overset{\longleftrightarrow}{}, X_n\}$ makes it possible to successively define the vector of the mean values and then the covariance matrix:

$$\{\mu(X)\} = \{\bar{x}_1, \bar{x}_2, \overset{\longleftrightarrow}{}, \bar{x}_n\}$$

[2.20]

$$\left[\{S(\{X\})\}\right] = \begin{bmatrix} Var\left(X_1\right) & Cov\left(X_1, X_2\right) & \cdots & Cov\left(X_1, X_n\right) \\ & Var\left(X_2\right) & \cdots & Cov\left(X_2, X_n\right) \\ & sym & & \ddots & \\ & & & & Var\left(X_n\right) \end{bmatrix}$$

[2.21]

CENTRAL LIMIT THEOREM.–

Let $\left(X_n\right)$ be a series of independent variables that follow the same probability law with the existence of a common expectation value m and a common variance σ^2. Then, the centered variable that is associated with the sum $S_n = X_1 + X^2 + \cdots\cdots\cdots + X_n$, namely $\dfrac{S_n - mx}{\sigma\sqrt{n}}$, converges to the reduced centered variable (hence of the Gaussian law $N(0,1)$).

2.2.5. *Normal probability functions*

The probability function that is most commonly used in scientific and engineering applications is the normal Gaussian distribution:

$$p(x) = \frac{1}{\sigma(x)\sqrt{2\Pi}} \exp(-\frac{x - \mu(x)^2}{2\sigma(x)^2}) \qquad [2.22]$$

A random variable is referred to as Gaussian (or normal) if its probability density function depends on the mean $\mu(x)$ and the standard deviation $\sigma(X)$ that fully characterizes any Gaussian random variable. A useful expression that is equivalent to equation [1.20] is obtained by defining a new random variable $Y = \dfrac{X - \sigma(X)}{\sigma(X)}$, which leads to the normalized expression of the Gaussian probability density:

$$p(y) = \frac{1}{\sqrt{2\Pi}} \exp(-\frac{y^2}{2}) \text{ with } \mu(Y) = 0 \text{ and } \sigma(Y) = 1. \qquad [2.23]$$

For a centered Gaussian distribution, the law-generating static moments are given by:

$$E[X^{2n+1}] = 0 \qquad [2.24]$$

$$E[X^{2n}] = \prod_{l=1}^{n} (2l - 1)(\sigma^2)^n \qquad [2.25]$$

with

$$\sigma^2 = E[X^2] = \text{var}(X) \qquad [2.26]$$

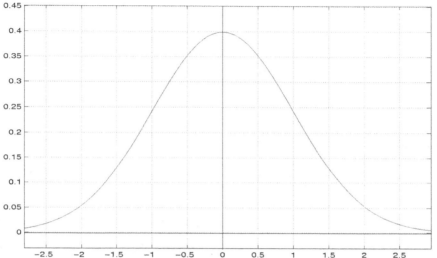

Figure 2.1. *Distribution of the normal law. For a color version of this figure, see www.iste.co.uk/dahoo/metrology1.zip*

The odd powers of the mathematical expectation are therefore null, and the even powers depend on the variance of the random variable.

The Gaussian probability density function has remarkable mathematical properties. It is worth mentioning that the linear functions of Gaussian random variables are also Gaussian.

2.2.6. *Uniform probability function*

The law of a uniform or rectangular probability distribution can be used to characterize random variables whose probability ranges within a specific interval and is proportional to the width of the interval.

Generally speaking, a variable "X" is referred to as a real uniform random variable on the interval [a, b] (where "a" and "b" are two real numbers such that $a \le b$) if its probability density is defined as:

$$f(x) = \begin{cases} \dfrac{1}{b-a} & \forall x \in [a,b] \\ \\ 0 & \forall x \notin [a,b] \end{cases} \qquad [2.27]$$

2.3. Random process and random field

The analysis of many phenomena requires the study of random variables that depend on several parameters.

A random process $\alpha(t, \theta)$ is defined as a parameterized family of random variables for which the index (or parameter) 't' belongs to a set "T" and θ belongs to the space of possible results of the random experiment.

In many engineering problems, the index is time; the fundamental intuitive notion is that of a random variable evolving in time. This is referred to as a random process.

In the presence of several parameters, there is a random field. Random variables are then distributed on a multidimensional space. They represent, for example, the random variation of the thickness of a plate. Therefore, the set of parameters contains spatial coordinates and may contain temporal coordinates.

For a specific value (sample or implementation) of the parameter(s), the value of the field consists of a random variable depending on the considered random phenomenon.

Hence, for a one-dimensional random process $\alpha(t, \theta)$, the result of the random experiment considered can be interpreted in four different ways [DI 00]:

– a family of functions of time (t and θ variable);

– a function of time (t variable, θ fixed);

– a random variable (t fixed, θ variable);

– a number (t fixed, θ fixed).

Let $\alpha(t,\theta)$ be a continuous stochastic field defined on the spatial–temporal coordinates 'x', associated with the "probability space of the random experiment". Since this field is defined in the domain Π of $\mathfrak{R}^d$, we have: $x = \{x_1, x_2, \rightleftarrows, x_d\}$.

A vector of random fields $\{\alpha(x,\theta)\}$ can also be defined; therefore, each component is a different random field, although defined by the identical sets of x :

$$\{\alpha(x,\theta)\} = \{\alpha_1(x,\theta), \alpha_2(x,\theta), \rightleftarrows, \alpha_p(x,\theta)\}$$

[2.28]

where 'P' is the number of random parameters.

Each vector of random fields $\{\alpha(x,\theta)\}$ is associated with a random parameter of the studied system, and its mean is defined by the spatial mathematical expectation:

$$\mu(\alpha_i(x)) = E\left[\alpha_i(x,\theta)\right] = \int_{-\infty}^{+\infty} \alpha_i(x,\theta) p_1(\alpha_i(x)) d\alpha_i(x)$$

[2.29]

Considering two random fields $\alpha_i(x,\theta)$ and $\alpha_j(x,\theta)$, the covariance can be defined by the following expression:

$$Cov(\alpha_i(x), \alpha_j(x)) = \int_{-\infty}^{+\infty} \int_{-\infty}^{+\infty} (\alpha_i(x,\theta) - \bar{\alpha}_i)(\alpha_j(x,\theta) - \bar{\alpha}_j)$$
$$p_2(\alpha_i(x), \alpha_j(x)) d\alpha_i(x) d\alpha_j(x)$$

[2.30]

with $i, j = 1, 2, \rightleftarrows, P$; $p_1(\alpha_i(x))$ being the probability density function of "$\alpha_i(x, \theta)$" and $p_2(\alpha_i(x), \alpha_j(x))$ being the joint probability density function of $\alpha_i(x,\theta)$ and $\alpha_j(x,\theta)$. The symbol $\overline{\alpha}$ indicates the mean $\mu(\alpha(x))$.

Similar to random variables, the mean and the covariance fully characterize a stochastic field of Gaussian type.

2.4. Mathematical formulation of the model

All the methods described apply to a physical system that has the following properties:

– environment with randomly fluctuating properties;

– random (external) excitation.

This can be expressed by the following equations:

$$\Gamma(x,\theta)u(x,\theta) = f(x,\theta) \qquad [2.31]$$

where Γ is a differential operator that can be rewritten as:

$$\Gamma(x,\theta) = L(x) + \Pi(\alpha(\theta), x) \qquad [2.32]$$

where L is the deterministic part of Γ and Π is the random part with zero mean.

Knowing Γ and f, the problem is to find the statistical properties of u.

The static properties of the solutions to the problem with eigenvalues associated with the following equation should also be determined:

$$\Gamma(x,\theta)\varphi(x,\theta) = \lambda(x,\theta)\varphi(x,\theta) \qquad [2.33]$$

where $\varphi(x,\theta)$ is the random eigenvector associated with the random eigenvalue $\lambda(x,\theta)$.

2.5. Reliability-based approach

The reliability-based approach involves uncertainty modeling. Depending on the methods used, uncertainties are modeled by random variables, stochastic fields or stochastic processes. These methods are used to study and analyze the variability of the system response and to minimize it.

The most common methods are the Monte Carlo method, the perturbations method and the polynomial chaos method [ELH 13].

2.5.1. *Monte Carlo method*

2.5.1.1. *Origin*

This mathematical tool was first used in 1930 in Fermi's research works on the characterization of new molecules. The Monte Carlo (MC) method was then applied in 1940 by von Neumann, Ulam and Fermi for simulations in atomic physics. The MC method is a powerful and very general mathematical tool. The computational power of current information technology opens a broad field of applications for this method.

2.5.1.2. *Principle*

The MC method is a computation technique that successively solves a deterministic system that incorporates uncertain parameters modeled by random variables.

The MC method is used when the problem is too complex to be solved analytically. It generates random drawings for all the uncertain parameters following their probability laws. The quality in terms of accuracy of random generators is very important. For each drawing, a set of parameters is obtained and a deterministic calculation is made.

2.5.1.3. *Advantages and drawbacks*

The main advantage of the MC method is essentially related to its applicability. This method potentially applies to all systems, irrespective of size or complexity. The results provided are statistically exact, meaning that their uncertainty decreases with the increase in the number of drawings. This uncertainty is defined for a given confidence level by the Bienaymé–Tchebychev inequality. A reasonable accuracy requires a large number of drawings. This is why the MC method is sometimes very costly in terms of calculation time, which is the major drawback of this method.

2.5.1.4. *Note*

Due to its simplicity, the MC method is applied in the field of engineering sciences. It is a powerful, but costly method. Its results are often used to validate new methods that are developed in fundamental research. It is applied in Chapter 9 of [DAH 16] for the characterization of carbon nanotubes.

2.5.2. *Perturbation method*

2.5.2.1. *Principle*

The perturbation method is another technique for studying the propagation of uncertainties on systems [KLE 92, ELH 13]. It involves the approximation of the functions of random variables by their Taylor expansion about their mean value. Depending on the order of Taylor expansion, the method is referred to as first order, second order or higher order. This method is subjected to the conditions of existence and validity of Taylor expansion, which limit its field of application to cases in which random variables have small dispersions about their mean value [ELH 13, GUE 15a].

The principle of the perturbation method involves the substitution, in the expression of the model response as a function of its parameters, of the random functions by their Taylor expansions. The same order terms are grouped and thus a system of equations is generated. The system is then solved successively, order by order, starting from zero. The mathematical formalism, as well as the general equations, can be found in the work of [ELH 13] and [GUE 15b].

2.5.2.2. *Applications*

There are numerous applications of the perturbation method. It can be used to solve various types of problems related to the propagation of uncertainties under both static and dynamic conditions, and linear and nonlinear conditions. However, good results are obtained only when the uncertain parameters have a low dispersion [ELH 13, GUE 15a].

In the context of vibrational behavior modeling and analysis of the dynamic behavior of systems, [GUE 15b] used the perturbation method for the study of the aerodynamic properties of elastic structures (stacked flat parts) subjected to several uncertain parameters (structural and geometric parameters). This work is the first published application of the stochastic finite element method combined with that of perturbations for the analysis of aerodynamic stability.

In another study, [ELH 13] presented a combination of the method of finite differences with that of perturbations in order to model the vibration problems in uncertain mechanical structures. The method developed is used, as an illustration, to determine the probabilistic moments of frequencies and eigenmodes of a beam whose Young modulus is random.

The second order is generally sufficient to obtain the first two moments with good accuracy. Muscolino [MUS 00] presented a method of dynamic analysis for linear systems with uncertain parameters and deterministic excitations. This method is an improvement of the first-order perturbation technique that sets limits when the dispersion of uncertain parameters is significant. The results obtained are compared to the reference results of the MC method and the second-order perturbation method. A strong correlation is observed between these results.

2.5.2.3. *Note*

The perturbation method relies on the simple principle by which all the random quantities are expressed by their Taylor expansions about their mean values. However, this method is difficult to implement particularly for systems with many degrees of freedom. Moreover, it can only be used when uncertain parameters have low dispersions around their mean values.

EXAMPLE 2.1.– Application of the perturbation method

The objective is to highlight the advantages of the perturbation method proposed by Muscolino in determining the beam response.

Let us consider a beam fixed at both ends, which is freely vibrating in the (Oxy) plane (Figure 2.1).

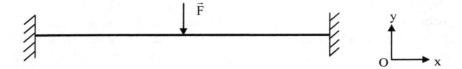

Figure 2.2. *Beam fixed at both ends*

Mass and rigidity matrices are given by:

$$[M] = \frac{m}{420}\begin{bmatrix} 156 & 22.1 & 54 & -13.1 \\ 22.1 & 4.1^2 & 13.1 & -3.1^2 \\ 54 & 13.1 & 156 & -22.1 \\ -13.1 & -3.1^2 & -22.1 & 4.1^2 \end{bmatrix} s$$

$$[K] = \frac{E.I}{l^3}\begin{bmatrix} 12 & 6.1 & -12 & 6.1 \\ 6.1 & 4.1^2 & -6.1 & 2.1^2 \\ -12 & -6.1 & 12 & -6.1 \\ 6.1 & 2.1^2 & -6.1 & 4.1^2 \end{bmatrix} \qquad [2.34]$$

The beam has a square cross-section of edge b that is considered as a Gaussian random variable.

The rigidity matrix [K] can be written in the following form:

$[K] = b^4 \cdot [A]$, where [A] is a deterministic matrix.

Similarly, the mass matrix [M] can be written as:

$[M] = b^2 \cdot [B]$, where [B] is a deterministic matrix.

Let us study the response of the beam to a force $F = 600\ sin(800t)$ acting at the center of the beam. The mean value and the standard deviation of the shift of the center of the beam are calculated using the second-order perturbation method and the new proposed method. The results obtained are compared to those obtained with the reference Monte Carlo technique using 10,000 drawings.

The results (Figures 2.2 and 2.3) show that the two perturbation methods yield the same results as the MC method.

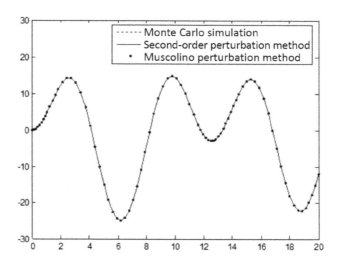

Figure 2.3. *Mean shift of the center of the beam*

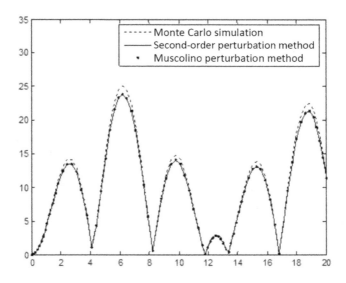

Figure 2.4. *Standard deviation of the center of the beam*

2.5.3. *Polynomial chaos method*

2.5.3.1. *Origins and principle*

The polynomial chaos method is a powerful mathematical tool developed by Wiener in his theory on homogeneous chaos [GUE 15a, GUE 15b]. This method formalizes a separation between the stochastic components of a random function and its deterministic components. Polynomial chaos is used to obtain a functional expression of a random response by decomposing its hazard on the basis of orthogonal polynomials.

Generally speaking, a field of second-order stochastic (finite variance) variables can be expressed by a series expansion of Hermite polynomials; orthogonal functions of certain random Gaussian and independent variables model the uncertainty. The deterministic part is modeled by the coefficients $\overline{x}_j$, known as stochastic modes, weighting the Hermite polynomial functions:

$$X(\xi) = \sum_{j=0}^{\infty} \overline{x}_j \phi_j(\xi) \qquad [2.35]$$

The family of polynomials Φ_i forms an optimal orthogonal basis and enables convergence in the sense of least squares of the expansion [ELH 13]. However, the rapidity of the convergence and the accuracy of the expansion into Hermite polynomials are no longer verified when dealing with non-Gaussian processes. In reality, the optimality of the Hermite basis in the case of Gaussian processes is due to the fact that the probability density function, which in this case is Gaussian, has a mathematical form that is similar to that of the weight function, which is associated with the scalar product defined in the basis. This principle can be generalized and used to establish a correspondence, referred to as the Askey scheme [ASK 85], between the families of orthogonal polynomials and the laws of probability. The notion of expansion into generalized chaos polynomials is then defined. An exponential convergence is thus proven and generalized to the case of random laws of probability (not necessarily Gaussian) [GHA 99].

2.5.3.2. *Note*

Polynomial chaos is a concept well suited to the modeling of random functions and processes. It is a tool for taking into account uncertainties and nonlinearities in system modeling and analysis. The numerical schemes for the implementation of polynomial chaos-based approaches differ by the way in which they use the model subject to the propagation of uncertainties. The advantage of the intrusive numerical scheme is that it requires only one calculation to determine the stochastic modes. However, this calculation becomes cumbersome when the original model has many uncertain parameters. The calculation complexity becomes more significant in the case of systems that have many degrees of freedom and are strongly nonlinear. This is because the original model is transformed through its projection on the polynomial chaos basis into a system of deterministic equations whose size and complexity depend significantly on the number of uncertain parameters and the number of degrees of freedom of the original model.

However, the non-intrusive scheme has a significant advantage, namely that it does not require modifications or manipulations of the original model. Several applications of this method can be found in [ELH 13].

2.6. Design of experiments method

2.6.1. *Principle*

The design of experiments method is used to implement or simplify, in terms of complexity and costs, an experimental protocol whose objective is to determine the parameters that have an impact on the performance of an industrial product. The purpose of this method is to obtain a design that is not sensitive to the variations of system parameters. By fixing the number of experiments to be conducted, this method makes it possible to determine the influence of several parameters on the responses of the system. Its efficiency when applied to a given system depends on the control of the values to be given to the parameters of the system and on the accuracy of the measurements of the corresponding responses. Several techniques rely on the notion of design of experiments. Various diagrams for building the design of experiments were described by Chatillon [CHA 05].

The use of the Taguchi design of experiments method significantly reduces the number of tests [TAG 86]. This method involves the intersection of two matrices of design of experiments: a control matrix representing adjustable factors and a noise matrix representing noise factors (uncertain parameters). The tests are conducted for combinations of factors that are identified in these matrices. Statistical quantities such as the mean value and the standard deviation of the response signal are measured. For the evaluation of results, the Taguchi method uses the signal-to-noise ratio and a loss function as the quality criterion. In the version of the method developed by [HUN 05], the notion of orthogonal columns is used for the simultaneous study of several design parameters, which reduces the minimal number of tests.

2.6.2. *Taguchi method*

This statistical method is used to define an experimental protocol whose objective is to render the main response of a system insensitive to various values of its parameters. It involves defining a set of experiments and the various sets of parameters to be used. The number of experiments to be conducted depends on the design parameters that are adjustable, the number of parameters that are variable (uncertain), potential dependences between these parameters and the effect of these parameters on the response (e.g. linear or nonlinear effect).

Considering the variability of many parameters, the Taguchi method makes it possible to optimizer the response of a system. This method originally used the signal-to-noise ratio as a quality index, combining the mean and the variance. The objective of the Taguchi method is to simplify the implementation of the design of experiments. It proposes a choice of matrices of experiments, tools for the choice of the most adapted table and recommendations for considering the interactions between design adjustment factors.

The collections of Taguchi tables serve to:

– choose the matrix of experiments to be developed depending on the number of factors, their modalities and their interactions;

– verify by means of linear graphs that the chosen table takes into account all the factors and their interactions and is a proper representation of the problem being considered;

– know by means of the interaction table the columns where the interactions that were neglected can be found.

EXAMPLE 2.2.– Application of design of experiments to robust design

The objective is to highlight the advantages of the design of experiments method in order to render the response of a system insensitive to the variations of its input variables. Let us consider a microcontroller component inserted in a rectangular printed circuit. The microcontroller component has 256 pins that are connected to the printed circuit by brazing joints. The circuit is fixed in a case made of aluminum alloy by five screws (one in each corner of the circuit and one in the central region of the card). FEM (finite element method) is applied to the card featuring a microcontroller component. The input parameters are the geometrical (position of the fifth screw, thickness of the printed circuit) and physical properties of the materials (printed circuit, brazing, pin, over-molding composite material of the electronic component). The response of the model is the strongest constraint among those exerted on the 256 brazing joints during a heat load.

To retain only the input variables that have a significant effect on the studied response, we focus on a screening design of experiments. When the number of variables is 35, a Plackett–Burman plan is developed. According to this plan, 15 parameters that have an effect on the response are retained.

To obtain the surface of the response, a Latin hypercube (LHS) design of experiments is used for the retained variables. The LHS design involving n tests is the design of experiments for which:

– each parameter has the same number of levels n (the higher this number, the finer the "mesh" and the higher the capacity of the adjusted model to apprehend the optima);

– each parameter has only one level.

When the parameters have different domains of variation, each parameter has n equally distributed levels between its minimum value and its maximum value. The LHS design is suitable for numerical tests due to its simplicity of implementation, as well as because numerical tests do not generate dispersion for the same test. The LHS design is suitable for the spatial interpolation method (Kriging) that is used to obtain the response surface.

The response signal considered is the constraint exerted on the brazing joint of the pin of the microcontroller component, which, among the 256 pins, has the most significant median constraint (a given pin of the component). Seventy percent of the tests of the LHS design of experiments are used to build the Kriging model. These tests are randomly drawn among the 200 tests. The remaining 30% are used to validate the predictive power of the model.

The resulting response surface makes it possible to approximate the constraint on the brazing joint, which is under the highest constraint among the total of 256. A total of 15 variables are considered in this model of response. To identify the variables with the highest influence on the constraint on the brazing joint, a global sensitivity analysis is conducted among these 15 variables, using the method of Sobol indices. A number of n simulations of input parameters are conducted, and the response is calculated for them (using the Kriging method). Then, in order to study the sensitivity of one of the parameters, a drawing is done once again for all the parameters except this one. This stage is repeated several times by bootstrap. Sobol indices for the studied parameter are then calculated, from the total variance and the variances related to the studied parameter. We obtain a statistical distribution for each index, which is represented as a boxplot, to estimate the confidence interval on the index value.

A total of 1,000 simulations are conducted and thus the input parameters are simulated and 100 indices are calculated. The parameters that have the highest influence on the response are listed below in the order of their importance:

– parameters X1 (COMP_Z): thickness of the brazing joint;

– parameter X4 (EX_SOLDER): Young's modulus of the brazing;

– parameter X5 (ALP_SOLDER): expansion coefficient of the brazing;

– parameter X14 (ALPX COMP): expansion coefficient of the component in the plane.

Once the influencing factors are identified, Monte Carlo simulations are conducted to determine the distribution of the constraint on the brazing joint, depending on the variations of the factors that influence it:

– each non-influencing factor is assigned the nominal value;

– for each influencing factor, a uniform law drawing is done in its field of variation;

– finally, the value of the constraint on the brazing joint is calculated using the Kriging model.

The previously described process is iterated a very large number of times (107), in order to obtain the distribution of the constraint on the brazing joint.

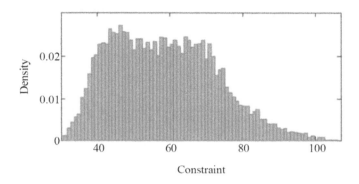

Figure 2.5. *Histogram of the constraint on the brazing joint. For a color version of this figure, see www.iste.co.uk/dahoo/metrology1.zip*

This distribution can be estimated by a parametric model, for example a Gaussian mixture model. The result is presented in Figure 2.5.

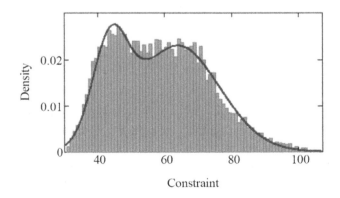

Figure 2.6. *Density of the constraint on the brazing joint. For a color version of this figure, see www.iste.co.uk/dahoo/metrology1.zip*

The objective is to adjust the level of the control factors in order to reduce the sensitivity of the system to the sources of variability (noise factors) and to harmonize the system response with its target (objective).

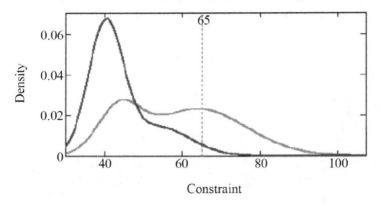

Figure 2.7. *Density of the initial constraint (red) and the optimized constraint (blue). For a color version of this figure, see www.iste.co.uk/dahoo/metrology1.zip*

Factor ALPX COMP (CTEX of the component) has a strong influence on the constraint (positive influence). This factor can be adjusted by the composite structure of the component coating material. Small values of ALPX COMP should be drawn to minimize the constraint. By initially reducing the range of variation of factor ALPX COMP to the interval [5;7], instead of [5;23], the mean value of the constraint is reduced, as well as its variability (Figure 2.6).

2.7. Set-based approach

The prerequisite of the above-described reliability-based approach is that the laws of probability governing the uncertain parameters are known. The advantage of the set-based approach is that it does not require models of the laws of probability of uncertainties. Two significant methods are found in the set-based approach: the first is based on interval arithmetic [MOO 66]; the second is based on the fuzzy logic formalism [ZAD 65].

2.7.1. *The interval method*

2.7.1.1. *Principle*

Interval calculation is rooted in the works of Moore [MOO 66]. This method relies on modeling uncertain parameters by intervals whose endpoints represent the minimal and maximal limits of the parameters. It can then be considered that the error between the model output and the system response is bounded and the endpoints are known. These endpoints take into account the measurement noise and modeling errors. The objective is to find a set of acceptable values instead of one value of the parameters, minimizing a convergence criterion. This method can be readily used. In contrast to probabilistic methods, no information is required on the nature of dispersions or on the way in which they evolve. However, it presents convergence difficulties.

2.7.1.2. *Interval arithmetic and stability analysis*

Interval arithmetic is applied to the analysis and stability of uncertain linear dynamic systems. Jaulin *et al.* [JAU 01] proposed a method for the characterization of all the values of uncertain parameters associating a stable dynamic behavior. For the study of stability, the solution is determined using the interval-based analysis according to the Routh criterion. Defining two sets A and B, stability analysis boils down to an inclusion problem. A is the admissible set of possible values of uncertain parameters, while B is the admissible set of values for which the system is stable. An algorithm based on the theory of intervals can be used to test the inclusion of A in B synonymous with a necessary and sufficient condition that proves the stability. The convergence of the algorithm is tested on numerical examples of systems.

EXAMPLE 2.3.– Interval method: vehicle suspensions

An example of application is a mass–spring–shock absorber system (Figure 2.7). This system is defined by the following equations:

$$\begin{cases} \dot{x}_s = v_s \\ \dot{x}_u = v_u \\ \dot{v}_s = -\frac{1}{m_s}\left(c_s\left(\dot{x}_s - \dot{x}_u\right) + k_s\left(x_s - x_u\right) + K_s\left(x_s - x_u\right)^3\right) \\ \dot{v}_u = -\frac{1}{m_u}\left(c_s\left(\dot{x}_s - \dot{x}_u\right) + k_s\left(x_s - x_u\right) + K_s\left(x_s - x_u\right)^3 - k_t\left(x_u - x_r\right) - K_t\left(x_u - x_r\right)^3\right) \end{cases} \quad [2.36]$$

where m_s and m_u are the masses, c is the damping coefficient, k_s and k_t are the linear rigidities, K_s and K_t are the cubic rigidities.

The initial conditions are:

$$[x_s, x_u, v_s, v_u]\big|_{t=0} = [0,0,0,0].$$

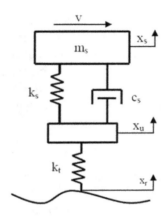

Figure 2.8. *Mass–spring–shock absorber system*

The parameters c_s, k_s and k_t are uncertain and modeled by intervals, as indicated in Table 2.1.

Parameters	m_s(kg)	C_s(Ns/m)	K_s(Ns/m)	K_t(Ns/m)	K_s(N/m³)	K_t(N/m³)
Mean value	375	1000	15000	200000	$1.5*10^6$	$2*10^7$
Interval	-	[900,1100]	[13500,16500]	[18,22]*10^4		

Table 2.1. *Parameters of the mass–spring–shock absorber system model*

The mean value of the shift is calculated using the interval method. The result (Figure 2.8) is compared to that obtained by the deterministic method. There is an agreement between the result of the interval method and that of the deterministic method.

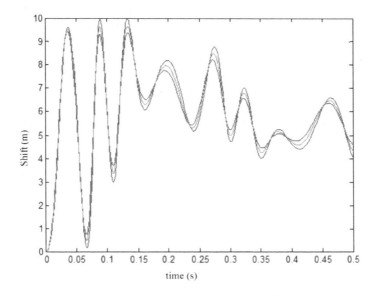

Figure 2.9. *Mean value of the shift xu(t) for the interval method (blue) and the deterministic method (red). For a color version of this figure, see www.iste.co.uk/dahoo/metrology1.zip*

2.7.1.3. *Conclusion*

Interval arithmetic can be used for modeling uncertainties only through their physical boundaries, which in most cases are identifiable and measurable. No information is required on the evolution of uncertainty in its interval.

2.7.2. *Fuzzy logic-based method*

2.7.2.1. *Principle*

The methods based on fuzzy logic were introduced for the interpretation and manipulation of uncertain data when no probabilistic or statistical information is available. These methods rely on the notion of a fuzzy set. An element of the fuzzy set, such as a model input value, has a degree of belonging to this set. This notion, which is formally known as the function of belonging, differs from the notion of probability. It defines an evasive quantitative measure of imperfect data. According to this definition, it is possible to establish a fuzzy logic associated with degrees of freedom, which, assigned to propositions, range from zero (false) to one (true) with all the possible graduations, leading to the following terms: a little, moderately, etc. Fuzzy logic can thus be applied to approximate reasoning.

EXAMPLE 2.4.–Application of the fuzzy logic-based method

Let us consider the application example of a two-dimensional frame under free vibration. This system, which is represented in Figure 2.9 in the (OXY) plane, consists of three beams of equal square section a. The only random parameter is the dimension a. The objective is to determine the stochastic shift of the horizontal beam of the frame during a given sinusoidal excitation ($F(t)$):

$$F(t) = 20 \, sin(80t)$$

The mean value and the standard deviation of the shift are calculated using the fuzzy logic-based method. The results (Figures 2.10 and 2.11) are compared to those of the reference method (Monte Carlo). It can be noted that the stochastic response of the frame (mean value and standard deviation of the shift of the beam) calculated by the fuzzy logic-based method is in agreement with the response calculated by the MC method.

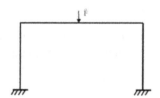

Figure 2.10. *Two-dimensional frame*

2.7.2.2. Conclusion

The fuzzy logic-based method for considering uncertainties can be used for dealing with vague, imprecise or linguistically described information. This uncertainty is described by form functions referred to as functions of belonging. The main advantage of this modeling is that it does not require statistical or probabilistic information. However, it is difficult to determine the function of belonging.

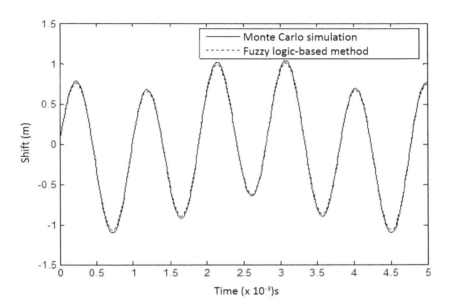

Figure 2.11. *Mean value of the shift as a function of time*

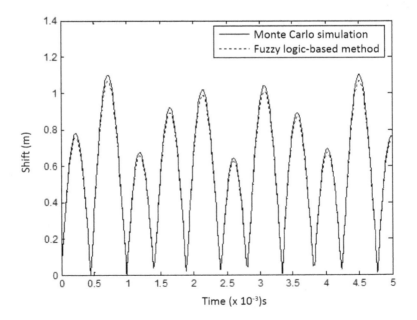

Figure 2.12. *Standard deviation of the shift as a function of time*

2.8. Analysis in terms of main components

The objective of this method is to determine the most significant components of a system depending on several variables. For example, a sample of N individuals characterized by P characters (or variables) is analyzed. The many relations between N and P will be analyzed.

In order to analyze the connection between two variables, their values are plotted on two orthogonal axes and the effect of these variables is analyzed using statistical tests. For three variables, this approach leads to a three-dimensional graphical representation. For four or more variables, the graphical representation is no longer possible. If the work is conducted on pairs or triplets of variables, complex interactions risk remaining hidden, which leads to the necessity of implementing a technique that extracts the most relevant information. This technique uses linear combinations of variables and is adapted to linear relations.

2.8.1. Description of the approach

Let X_1, X_2,... X_p be the initial quantitative and centered variables (of zero mean). A new variable Y_1 is determined as a linear combination of X_i:

$$Y_1 = c_1 X_1 + c_2 X_2 + ... + c_p X_p \qquad [2.37]$$

where c_1, c_2, ..., c_p are constants to be determined, such that Y_1 has a maximum variance with the following constraint:

$$c_1^2 + c_2^2 + ... + c_p^2 = 1 \qquad [2.38]$$

Among all the linear combinations of X_i, the one that generates the least possible destruction of information has the widest possible dispersion. If Y_1 has zero dispersion, Y_1 is a constant. The problem to be solved is to determine the constants c normed to 1 for which the variance of Y_1 is maximal. The constants c (and therefore Y_1) can be determined. The variable Y_1 is referred to as the main component and V_1 is its variance.

In general, Y_1 does not exhaust all of the variance of the original variables. Then, a second variable Y_2 is determined, not correlated with Y_1, of maximum variance, as a linear combination of X_i:

$$Y_2 = c_{12}X_1 + c_{22}X_2 + \dots + c_{p2}X_p \qquad [2.39]$$

where c_{12}, c_{22}, ..., c_{p2} are constants to be determined under the normalization constraint:

$$c_{12}^2 + c_{22}^2 + \dots + c_{p2}^2 = 1 \qquad [2.40]$$

NOTE.–

It is possible to replace c_{11} with c_1 and c_{21} with c_2, etc. It can then be proven that the constants c (and hence Y_2) are uniquely determined.

Let V_2 be the variance of the new variable Y_2. By construction, we have: $V_1 \geq V_2$.

Y_2 is referred to as the second main component. New variables Y_3, $Y_4, \dots, Y_p$ can be built in a similar manner. These variables are not correlated with the previous ones, of maximum variance (the normalization condition being related to the coefficients of the linear combination).

Let V_3, $V_4, \dots, V_p$ be the variances of these new variables. We then have:

$$V_3 \geq V_4 \geq V_5 \dots \geq V_p \qquad [2.41]$$

2.8.2. Mathematical basis

Determining constants c (or those of Y) amounts to determining eigenvalues. The various c are the coordinates of (normed) eigenvectors of the variance–covariance matrix of the initial variables X. The variances V_1, $V_2, \dots, V_p$ are the associated eigenvalues. The various cited properties (existence and uniqueness) can be deduced from them. If the rth variance V_{r+1} is very small, variables Y_{r+1}, $Y_{r+2}, \dots, Y_p$ are approximately constant for all the individuals. Keeping only the main components, Y_1, $Y_2, \dots, Y_p$ is therefore natural. In practice, V_{r+1} is estimated to be small if:

$$\frac{(V_1 + V_2 + \dots + V_r)}{(V_1 + V_2 + \dots + V_p)} \approx 90\% \qquad [2.42]$$

In the best case scenario, three main components are sufficient. The P correlated variables are then reduced to three uncorrelated variables that can easily be graphically represented.

2.8.3. *Interpretation of results*

The initial objective of extracting the most relevant information is generally reached. The number of variables is smaller (main components). They are uncorrelated and the individuals can easily be graphically represented without much distortion. There are two approaches: one is based on variables and the other on individuals.

2.8.3.1. *Method based on variables*

The correlation between the main components and the original variables is determined. If only the first r main components Y_1, Y_2,...,Y_r are kept, we obtain r.P correlation coefficients to be calculated: the correlation of Y_1 with X_1, X_2,...,X_p, that of Y_2 with X_1, X_2,...,X_p, ..., that of Y_p with X_1, X_2,...,X_p. The main components are interpreted on the basis of the observed values of these coefficients.

2.8.3.2. *Method based on individuals*

The main components can be interpreted by means of the position of the individuals with respect to the main components. The individuals whose contributions related to the concerned axes are too small are considered poorly represented. The position of the individuals in the planes formed by the components can be interpreted.

2.9. Applications

2.9.1. *Rod mesh*

This example takes into account a rod structure (rods are identical to the rod dealt with in the previous example) such as the plane mesh presented in Figure 2.13. The chords are articulated to nodes and contribute to the overall stiffness only by their extension. Structural stiffness and mass matrices can be easily built using the rod finite element [RIT 08].

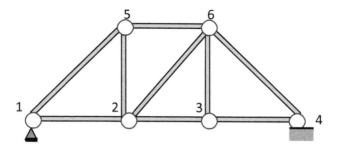

Figure 2.13. *Mesh structure*

The mass and stiffness matrices of the structure are according to Géradain [GÉR 96]:

$$[K]\frac{L}{ES} = \begin{bmatrix} 2+\frac{1}{2\sqrt{2}} & & & & & & & & & \\ \frac{1}{2\sqrt{2}} & 1+\frac{1}{2\sqrt{2}} & & & & & & & & \\ -1 & 0 & 2 & & & & & & & \\ 0 & 0 & 0 & 1 & & & \text{sym.} & & & \\ 0 & 0 & -1 & 0 & 1+\frac{1}{2\sqrt{2}} & & & & & \\ 0 & 0 & 0 & 0 & 0 & 1+\frac{1}{2\sqrt{2}} & & & & \\ 0 & -1 & 0 & 0 & 0 & \frac{1}{2\sqrt{2}} & 1+\frac{1}{\sqrt{2}} & & & \\ \frac{-1}{2\sqrt{2}} & \frac{-1}{2\sqrt{2}} & 0 & 0 & \frac{-1}{2\sqrt{2}} & -1 & 0 & 1+\frac{1}{\sqrt{2}} & & \\ \frac{-1}{2\sqrt{2}} & \frac{-1}{2\sqrt{2}} & 0 & -1 & \frac{1}{2\sqrt{2}} & 0 & 0 & 0 & 1+\frac{1}{\sqrt{2}} \end{bmatrix}$$

$$[M]\frac{6}{ml} = \begin{bmatrix} 2(3+\sqrt{2}) \\ 0 & 2(3+\sqrt{2}) \\ 1 & 0 & 6 & & & & sym. \\ 0 & 1 & 0 & 6 \\ 0 & 0 & 1 & 0 & 2(1+\sqrt{2}) \\ 1 & 0 & 0 & 0 & 0 & 2(2+\sqrt{2}) \\ 0 & 1 & 0 & 0 & 0 & 0 & 2(2+\sqrt{2}) \\ \sqrt{2} & 0 & 1 & 0 & \sqrt{2} & 1 & 0 & 4(1+\sqrt{2}) \\ 0 & \sqrt{2} & 0 & 1 & 0 & 0 & 1 & 0 & 4(1+\sqrt{2}) \end{bmatrix}$$

The dynamic problem can be written as:

$$\left(\frac{ES}{L}(1+i\eta)[K] - w^2 \frac{mL}{6}[M] \right)\{H\} = \{F\} \tag{2.43}$$

with the Gaussian E of mean E_0 and standard deviation σ_E. E can be expressed as:

$$E = E_0 + \sigma_E \zeta \tag{2.44}$$

where ζ is a reduced centered normal variable.

Then:

$$\left(\frac{S}{L}(E_0 + \sigma_E \zeta)[K](1+i\eta) - w^2 \frac{mL}{6}[M] \right)\{H\} = \{F\} \tag{2.45}$$

Equation [1.46] can be solved by expanding $\{H\}$ on a second-order chaos 2.

Figure 4.15 presents the mean value of the magnitude of the transfer function; the numerical values considered are a standard deviation of 1% and a damping of 4%. The calculations required for various methods are presented below. It can be easily noted that the time needed for the projection method is very short compared to the direct simulation method, knowing that the frequency domain of interest is divided into 401 points:

Direct Monte Carlo simulation, 2,000 drawings: 70.7400 seconds.

Projection on a second-order polynomial chaos: 0.8200 seconds.

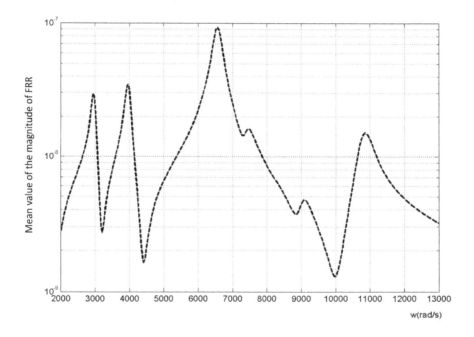

Figure 2.14. *Mean value of the module of the transfer function*

Figure 2.14 presents the mean value of the module of the transfer function located in (1,1) calculated by an expansion on a second-order polynomial chaos and a Monte Carlo simulation.

2.9.2. *Example of a linear oscillator*

Let us consider the application example of the forced response of a linear oscillator shown in Figure 2.8, which is governed by the following equation:

$$\ddot{x} + 2\xi\Omega\dot{x} + \Omega^2 x = f\sin(\omega t) \qquad [2.46]$$

In this example, the equivalent viscous damping coefficient ξ and the undamped eigenfrequency Ω are considered as uncertain parameters that are described by truncated Gaussian probability densities that are statistically independent. Truncation is introduced because the values of uncertain parameters are physically bounded. It is particularly important to have positive values of Ω and ξ. For Gaussian variables, the truncation is at about three times the standard deviation about the mean value.

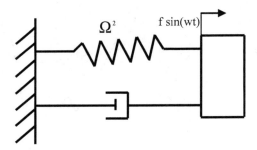

Figure 2.15. *Linear oscillator under study*

The equivalent viscous damping and the undamped eigenfrequency can be rewritten as:

$$\xi = \xi_0 + \varepsilon\xi_1 \qquad [2.47]$$

$$\Omega = \Omega_0 + \varepsilon\Omega_1 \qquad [2.48]$$

The parameter ε is a perturbation parameter that is assumed to be small.

The index 0 represents the mean value of the random quantity, while the index 1 represents the centered Gaussian random fluctuation.

Let us consider the variability of the module of the forced response during motion in the case of uncertain parameters.

The equivalent viscous damping coefficient ξ and the undamped eigenfrequency Ω have, respectively, mean values of 5% and 1 rad/s and standard deviations of 5% and 0.05 rad/s.

The standard deviation of the response was calculated by the Taguchi method with nine points of discretization for each random variable. The result is compared to that obtained by the MC method, using 10,000 simulations.

The result is shown in Figure 2.15. The random result shows a widening of the peak of deterministic resonance and a sensitive decrease of the resonance level. It is worth noting that the Taguchi method with nine points behaves very well compared to the Monte Carlo simulation.

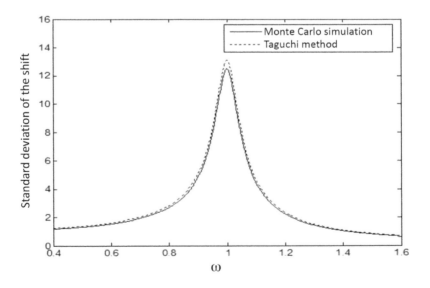

Figure 2.16. *Standard deviation of the module of the shift response*

2.10. Conclusion

This chapter presents several statistical methods for assessing the effect of uncertainties in the system response according to various approaches. If experimental data are described by probability laws, then the probabilistic approach is recommended. If the data are within an interval without any information on their variation, the interval-based algebraic approach is more convenient, but a problem of convergence may sometimes remain. When no probabilistic or statistical information are available, and there are unexpected ranges, fuzzy logic methods are more appropriate. There are several design of experiments methods that strongly reduce the dimensions of the problem being considered. The analysis of the main components can be used to determine one or two components with the strongest influence on the system with respect to given indicators.

3

Electromagnetic Waves and their Applications

3.1. Introduction

Electromagnetic waves can be used to study a physical–chemical (molecule, material, atmosphere, environment) or a technological system (nano-systems, MEMS) by analyzing their interactions. When the range of the electromagnetic waves is between UV and infrared, light–matter interaction occurs. The principle on which this is based is similar in other ranges of the electromagnetic spectrum. In general, the electromagnetic wave incident on an object is generated in a controlled state by the use of an adapted light source and filter. A detection system composed of adapted sensors and filters and embedded electronics is used for measuring the modifications undergone by the electromagnetic wave with respect to its initial state. By unraveling the physical processes by which the characteristics of the reflected, transmitted or scattered electromagnetic wave are modified with respect to the incident wave, the properties of the system can be assessed.

Depending on the device used to prepare the incident wave, and on the detection system, the volume of matter or material can be probed on the nanometer scale [DAH 16] (Figure 3.1). To assess the current state of these test techniques, a synthesis of the description of light phenomena is presented, starting with the ancient Greeks and up to modern times. In the first models of light properties, the ancient Greeks relied on the particle nature of light. For example, using geometrical optics (600 BC), Thales of Miletus measured the height of the pyramid of Cheops based on the length of the shadow of a rod, with the sun as a source of light. Euclid (300 BC) and Ptolemy (280 BC), respectively, described in their books ("Optica" and "Optics") the properties of light in the context of geometrical optics. From 1620 onwards, Snell and Descartes represented light by rays that form light beams that travel in a straight line in a homogeneous medium and follow the Snell–Descartes laws of reflection and refraction. Fermat principle (1657) on the time extremum of the optical path of a light ray led likewise to these results [BRU 65, HEC 05, MEI 15].

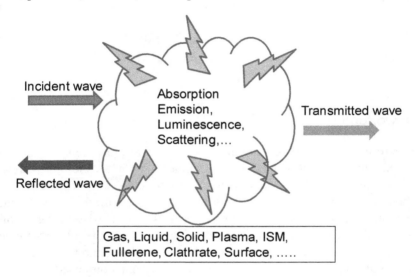

Figure 3.1. *Light–matter interaction. For a color version of this figure, see www.iste.co.uk/dahoo/metrology1.zip*

But light passing through a very small aperture undergoes diffraction. This phenomenon contradicts the theory of geometrical optics: the ray cannot be localized. In the same century, the wave-like properties were established on the basis of three postulates and the qualitative description of C. Huygens (1690). On this basis, A. Fresnel (1815) built a mathematical model for light propagation based on the existence of wave fronts. A point-source emits spherical wave fronts. Each point of the wave fronts behaves like a secondary point source emitting spherical waves in all directions. The secondary waves interfere so that the envelope of all the secondary wave fronts constitutes the new wave front. Due to this approach, light interference (T. Young experiments, 1801) and diffraction (F.M. Grimaldi experiments, 1664) could be interpreted. Furthermore, A. Fresnel assumed that the light wave is perpendicular to its direction of propagation (suggestion of A-M. Ampère, 1816). He proved that, when the incidence angle is non-zero, the coefficients of the reflection of light at the interface between two media have different expressions for an S wave (when the vibration is perpendicular on the plane of incidence) and a P wave (when light vibration is in the plane of incidence). Transverse polarization of light was thus formulated.

The approach developed by Huygens and Fresnel explains diffraction and interference phenomena. Approximately two centuries later (1865), James Clark Maxwell [MAX 54] unified the laws of electrostatics, magnetism and induction in a system of equations, simplified by Heaviside, which are known as Maxwell equations, in which electric and magnetic phenomena are considered in their field form. His equations led to postulating the existence of electromagnetic waves traveling with the speed of light ($3 \ 10^8$ m/s or more exactly $2.99,729,458 \times 10^8$ m/s). Visible light covers only one part of the spectrum of frequencies from 500 kHz to 10^{14} GHz. The works of Maxwell proved the vector nature of light. The electric field that is perpendicular to the direction of propagation can be located by two different directions known as the polarization states of light.

The Newtonian particle theory of light does not explain interference and diffraction phenomena, but the corpuscular approach developed by Einstein (1905), according to which light is constituted of grains of energy, provides an interpretation of the photoelectric effect. Light interacts with matter in the form of energy quantum ($E = h\nu$), where ν is the frequency associated with the color of light. In the case of the photoelectric effect, induced absorption takes place. Modeling a black body as a source that radiates energy in discrete form, Planck solved the problem of the ultraviolet catastrophe of the black body emission. In 1915, Einstein explained the black body emission by introducing a symmetrical emission process in parallel with the absorption process that occurs in discrete form. Light can thus be considered as a wave or as a particle [BRO 68]. These approaches have been developed in specialized papers or books. The differences between classical sources (incoherent light) and quantum sources (coherent light) can be established from statistical theories of light. Notions of quantum mechanics are required to understand the particle nature of light in the form of photons [MES 64]

[COH 73] such as the Glauber formulation, for example. To describe the state of wave polarization in a simplified form, a gauge is used to obtain the equation of the propagation of the electromagnetic wave in the transverse representation. This gauge choice makes it possible to express the electromagnetic field as a sum of independent oscillators (using creation and annihilation operators). This naturally leads to a quantum description of light in terms of photons, grains of light proposed by Einstein to interpret the photoelectric effect. It can be shown that energy is associated with a frequency and that light has a discrete character. Finally, Glauber's approach connects the classical (continuous) and the quantum (discrete) descriptions and attributes a physical reality to coherent states with photons distributed according to a Poisson law.

Nowadays, the wave–particle duality of light can be verified experimentally. The non-locality phenomenon was evidenced in the experiments conducted by Alain Aspect on quantum entanglement [ASP 82].

This chapter is dedicated to various applications of electromagnetic waves. After having recalled the characteristics of these waves, applications are presented: the energy carried by a monochromatic plane wave, a rectangular waveguide, microwave antennas, the study of a wire antenna and antenna networks. These applications help understand the principles on which the fifth generation (5G) of mobile telecommunication systems are based.

3.2. Characteristics of the energy carried by an electromagnetic wave

Electromagnetic waves carry energy. If $\vec{E}$ is the electric field and $\overline{B}$ is the magnetic field of the electromagnetic wave, the direction of the wave is the direction of the vector $\vec{E} \times \vec{B}$.

The Poynting vector $\vec{S}$ is defined by:

$$\vec{S} = \frac{1}{0}\vec{E}\times\vec{B}$$

$\vec{S}$ is expressed in joule per second per unit surface $\left[J / \left(s.m^2 \right) \right]$.

The norm of the Poynting vector represents the instantaneous power carried by the electromagnetic wave through the unit surface. The Poynting vector is perpendicular to the wave direction of propagation.

The Poynting vector can also be written in the following integral form (though denoted by the same letter, the energy E and the electric field $\vec{E}$ should not be confused):

$$P = \frac{dE}{dt} = \frac{1}{\mu_0} \oint_S \left\| \vec{E} \times \vec{B} \right\| dS = \frac{1}{\mu_0} \oint_S \left(\vec{E} \times \vec{B} \right) \circ d\vec{S} = c^2 \varepsilon_0 \oint_S \left(\vec{E} \times \vec{B} \right) \circ \vec{n} dS$$

where $\vec{n}$ is the unit vector perpendicular to dS.

For a plane electromagnetic wave, the norm of the Poynting vector is:

$$S(x,t) = \frac{E(x,t) \cdot B(x,t) \cdot \sin(\pi/2)}{\mu_0} = \frac{\hat{E} \cdot \hat{B}}{\mu_0} \cdot \sin^2(kx - \omega t) = \frac{\hat{E}^2}{\mu_0 \cdot c} \cdot \sin^2(kx - \omega t)$$

This quantity varies in time and space. In a given position, its average value is $\sin^2(\ )$ averaged over period T:

As

$$\frac{1}{T} \int_0^T \sin^2(\omega \cdot t) dt = \frac{1}{T} \left[\frac{1}{2} \int_0^T dt - \frac{1}{2} \int_0^T \cos(2\omega \cdot t) dt \right]$$

$$= \frac{1}{T} \left[\frac{T}{2} - \frac{1}{2} \left[\frac{1}{2\omega} \cdot \sin(2\omega \cdot t) \right]_0^T \right] = \frac{1}{2}$$

Hence:

$$\forall x \Rightarrow \overline{S} = \frac{\hat{E} \cdot \hat{B}}{2\mu_0} = \frac{\hat{E}^2}{2\mu_0 c} = \frac{c\hat{B}^2}{2\mu_0}$$

The average value of the Poynting vector of a plane electromagnetic wave is a constant that depends on neither position nor time.

The electric energy E_g of an ideal plate capacitor C (of section S, length L and electric permittivity ε) at a potential U generating plane electromagnetic waves is:

$$E_g = \frac{1}{2}C \cdot U = \frac{1}{2}\left(\frac{S \cdot \varepsilon}{L}\right) \cdot \left(\hat{E}^2 \cdot L^2\right) = \frac{1}{2} \cdot S \cdot L \cdot \varepsilon \cdot \hat{E}^2 = \frac{1}{2} \cdot V \cdot \varepsilon \cdot \hat{E}^2$$

$$\Rightarrow \frac{E_g}{V} = \frac{1}{2} \cdot \varepsilon \cdot \hat{E}^2$$

The energy density is:

$$u_g = \frac{1}{2} \cdot \varepsilon \cdot \hat{E}^2$$

which leads us to:

$$u_g = \frac{1}{2\mu}\hat{B}^2$$

and the total energy u_{tot} carried by the electromagnetic wave in this specific case is:

$$u_{tot} = \frac{1}{2}\varepsilon\hat{E}^2 + \frac{1}{2\mu}\hat{B}^2 = \frac{1}{2}\varepsilon\hat{E}^2 + \frac{1}{2\mu c^2}\hat{E}^2 = \varepsilon\hat{E}^2 = \frac{1}{\mu}\hat{B}^2$$

The electric energy density of an electromagnetic wave is equal to its magnetic energy density.

Following this result, "the (average) intensity I of an electromagnetic wave" can be defined by the average value of its Poynting vector:

$$I = \overline{S}$$

It is therefore the average power carried by the wave on the unit surface. But given the above proven time average expression of the Poynting vector, it can be written as:

$$I = \overline{S} = \frac{\hat{E} \cdot \hat{B}}{2\mu_0}$$

Using the relation between energy and momentum:

$$p = \frac{E_{tot}}{c}$$

The momentum density of the electromagnetic wave is written as:

$$p = \frac{u_{tot}}{c} = \frac{\varepsilon \hat{E}^2}{c} = \frac{\hat{B}^2 c}{\mu}$$

As the direction of $\vec{E} \times \vec{B}$ is perpendicular to the wave front and therefore identical with the wave propagation direction, its module is:

$$\left\| \vec{E} \times \vec{B} \right\| = \hat{E} \cdot \hat{B} = \frac{1}{c} \hat{E}^2 = c\hat{B}^2$$

Therefore, the momentum density of the electromagnetic wave can be written as:

$$p = \varepsilon \left\| \vec{E} \times \vec{B} \right\| = \frac{1}{\mu c^2} \left\| \vec{E} \times \vec{B} \right\|$$

As the direction of the momentum is that of the propagation, the vector form of the relation is expressed as:

$$\vec{p} = \varepsilon \vec{E} \times \vec{B} = \frac{1}{\mu c^2} \vec{E} \times \vec{B}$$

If an electromagnetic wave has a momentum, it also has a density of the angular momentum. The angular momentum per unit volume is then:

$$\vec{b} = \vec{r} \times \vec{p} = \varepsilon \vec{r} \times \left(\vec{E} \times \vec{B} \right) = \frac{1}{\mu c^2} \vec{r} \times \left(\vec{E} \times \vec{B} \right)$$

Therefore, an electromagnetic wave carries a momentum and an angular momentum besides energy.

This is not surprising. An electromagnetic interaction between two electrical charges involves energy and momentum exchange between the charges. This takes place through the electromagnetic field that carries the exchanged density of energy and momentum.

3.3. The energy of a plane monochromatic electromagnetic wave

A plane monochromatic electromagnetic wave travels in a vacuum along direction Oz. The electric components of this wave in the directions x, y and z are expressed by:

$$E_x = E_{ox}\, cos\, (kz - \omega t + \Phi_1)$$

$$E_y = E_{oy}\, cos\, (kz - \omega t + \Phi_2)$$

$$E_z = 0$$

1) Determine the curve described by the tip of the electric field vector in the following cases:

$$- \Phi_1 = \Phi_2$$

$$- E_{ox} = E_{oy}\, \text{and}\, \Phi_1 - \Phi_2 = \pi/2$$

2) Determine the dispersion relation.

3) Determine the relation between E and B. Write the explicit expression of $B\,(x, y, z, t)$.

4) Calculate the electromagnetic energy density $u\,(r, t)$ at any point.

5) Determine in various ways the Poynting vector $R\,(r, t)$, equal to the current density $j_u\,(r, t)$ of the electromagnetic energy:

a) using a general expression of R as a function of E and B;

b) using the relation $j_u = u\, v_u$;

c) using the conservation equation that reflects the conservation of the electromagnetic energy in a vacuum. The direction of R should first be determined by symmetry.

6) Calculate the average power through the unit surface S that is normal to the propagation direction.

3.3.1. *Answer to question 1*

3.3.1.1. *First case*

Since $\Phi_1 = \Phi_2$

$$\frac{E_x}{E_{0x}} = \frac{E_y}{E_{0y}}$$

For example, in the plane z=0, this means that:

$\forall t, \tan \theta = \dfrac{E_{0y}.\cos(\omega t)}{E_{0x}.\cos(\omega t)} = \dfrac{E_{0y}}{E_{0x}} = \dfrac{E_y}{E_x}$, hence $\tan \theta$ is constant.

As angle θ is constant, the tip of the electric field vector describes a straight line, which corresponds to a state of rectilinear polarization. The same is applicable to any plane z = constant.

3.3.1.2. *Second case*

Since $E_{ox,} = E_{oy}$ and $\Phi_1 - \Phi_2 = \pi/2$, the components of the electric field which are expressed by:

$$\begin{pmatrix} E_x = E_{0x} \cos(kz - \omega t + \Phi_1) \\ E_y = E_{0x} \cos\left(kz - \omega t + \Phi_1 - \dfrac{\pi}{2}\right) = E_{0x} \sin(kz - \omega t + \Phi_1) \\ E_z = 0 \end{pmatrix}$$

verify the following equation:

$\left(\dfrac{E_x}{E_{0x}}\right)^2 + \left(\dfrac{E_y}{E_{0x}}\right)^2 = 1$, which is the equation of a circle of radius E_{0x}.

In this case, the tip of the electric field vector describes a circle; the wave is circularly polarized.

3.3.2. *Answer to question 2*

The dispersion equation is established using the wave equation that links the second-order time derivative of the wave to its second-order space derivative:

$$k^2 = -\mu_0 \times \varepsilon_0 \times \omega^2 = 0$$

$$k^2 = \mu_0 \times \varepsilon_0 \times \omega^2$$

3.3.3. *Answer to question 3*

Using the relation: $\overrightarrow{rot}\ \vec{E} = \dfrac{-\partial \vec{B}}{\partial t}$ in the case of a plane wave, the nabla operators and the partial derivative with respect to time can be replaced by multiplication by $j\,k$ and $-j\,\omega$.

Considering the case $\Phi_1 = \Phi_2 = 0$

$$\begin{pmatrix} E_x = E_{0x}.\cos\ (kz - \omega t + \Phi_1) \\ E_y = E_{0y}.\cos\ (kz - \omega t + \Phi_2) \\ E_z = 0 \end{pmatrix} = \begin{pmatrix} E_x = E_{0x}.\cos\ (kz - \omega t) \\ E_y = E_{0y}.\cos\ (kz - \omega t) \\ E_z = 0 \end{pmatrix}$$

$$j\vec{k} \wedge \begin{pmatrix} E_x = E_{0x}.\exp(j(kz - \omega t)) \\ E_y = E_{0y}.\exp(j(kz - \omega t)) \\ E_z = 0 \end{pmatrix} = -j\omega \begin{pmatrix} B_x \\ B_y \\ B_z \end{pmatrix}$$

$$j\begin{pmatrix} 0 \\ 0 \\ k \end{pmatrix} \wedge \begin{pmatrix} E_x = E_{0x}.\exp(j(kz - \omega t)) \\ E_y = E_{0y}.\exp(j(kz - \omega t)) \\ E_z = 0 \end{pmatrix} = \begin{pmatrix} -jkE_{0y}.\exp(j(kz - \omega t)) \\ +jkE_{0x}.\exp(j(kz - \omega t)) \\ 0 \end{pmatrix}$$

$$-j\omega \begin{pmatrix} B_x \\ B_y \\ B_z \end{pmatrix} = \begin{pmatrix} -jkE_{0y}.\exp(j(kz-\omega t)) \\ +jkE_{0x}.\exp(j(kz-\omega t)) \\ 0 \end{pmatrix}$$

$$\begin{pmatrix} B_x \\ B_y \\ B_z \end{pmatrix} = \begin{pmatrix} \dfrac{k}{\omega}E_{0y}.\exp(j(kz-\omega t)) \\ -\dfrac{k}{\omega}E_{0x}.\exp(j(kz-\omega t)) \\ 0 \end{pmatrix} = \begin{pmatrix} \dfrac{k}{\omega}E_{0y}.\cos((kz-\omega t)) \\ -\dfrac{k}{\omega}E_{0x}.\cos((kz-\omega t)) \\ 0 \end{pmatrix}$$

3.3.4. *Answer to question 4*

The electromagnetic energy density is given by the formula:

$$u = \frac{\varepsilon_0}{2}\vec{E}^2 + \frac{1}{2\mu_0}\vec{B}^2$$

Hence, the following expression is based on the expressions of E and B:

$$du = \frac{\partial u}{\partial t}dt + \frac{\partial u}{\partial z}dz$$

For a constant energy density, $du = 0$, hence:

$$U_\varepsilon = \frac{-\dfrac{\partial u}{\partial t}}{\dfrac{\partial u}{\partial z}}$$

3.3.5. *Answer to question 5*

The general expression of R as a function of E and B is applied.

For the calculation of the Poynting vector: $\vec{R} = \dfrac{\vec{E} \wedge \vec{B}}{\mu_0}$; the real part of E and B must be considered.

If E and B are perpendicular, it can be readily written:

$$\left|\vec{R}\right| = \frac{\left|\vec{E}\right| \cdot \left|\vec{B}\right|}{\mu_0} = \frac{E^2}{c\mu_0}$$

The energy density is given by u, with $\mu_0 \varepsilon_0 c^2 = 1$.

Applying the relation $j_u = u\, v_u$, the following relation is obtained:

$$\left|\vec{j_u}\right| = uc = c\varepsilon_0 E^2 = \frac{E^2}{\mu_0 . c} = \left|\vec{R}\right|$$

where "c" is the speed of propagation of light.

The conservation equation, which reflects the conservation of electromagnetic energy in a vacuum, is used. This method requires the prior specification by symmetry of the direction of R.

By analogy with the charge conservation equation, the electromagnetic energy conservation equation can be written straightforward. Using the charge conservation equation, the electromagnetic energy conservation equation can thus be obtained.

From the charge conservation equation: $\dfrac{\partial}{\partial t}\rho + div\, j = 0$ with $\vec{j} = \rho\vec{V}$.

The electromagnetic energy conservation equation can be written as:

$$\frac{\partial}{\partial t}U + div\, \vec{R} = 0$$

Since:

$$div \, \vec{R} = \frac{\partial}{\partial z} R$$

and

$$\frac{\partial}{\partial t} U = 2\varepsilon_0 \omega (E_{ox}^2 \cos(\phi_x) \sin(\phi_x) + E_{oy}^2 \cos(\phi_y) \sin(\phi_y))$$

$$\phi_x = kz - \omega t + \varphi_1 \;\; and \;\; \phi_y = kz - \omega t + \varphi_2$$

and

$$R = \int 2\varepsilon_0 \omega (E_{ox}^2 \cos(\phi_x) \sin(\phi_x) dz + \int E_{oy}^2 \cos(\phi_y) \sin(\phi_y)) dz$$

$$\phi_x = kz - \omega t + \varphi_1 \;\; and \;\; \phi_y = kz - \omega t + \varphi_2$$

By integration, it can be obtained:

$$\left| \vec{R} \right| = \varepsilon_0 c \, \vec{E}^2 = \frac{\vec{E}^2}{c \mu_0}$$

3.3.6. *Answer to question 6*

The power is given by: $P(t) = \iint \vec{R}.\vec{dS} = R.S$

The average power is equal to: $\langle P(t) \rangle = \frac{1}{2} \frac{\varepsilon_0}{\mu_0} E^2.S$

3.4. Rectangular waveguide as a high-pass frequency filter

The purpose of this exercise is to show that a given volume of finite dimensions cannot contain all the photons of the electromagnetic spectrum, in other words, all the possible wavelengths (respectively, frequencies).

The speed of light in a vacuum, (c), depends on the properties of the vacuum, characterized by its electric permittivity $\varepsilon_0 = 8.86 \times 10^{-12}$ F.m^{-1} and its magnetic permeability $\mu_0 = 4\pi 10^{-7}$ H m^{-1} and is given by the well-known relation: $\varepsilon_0 \mu_0 c^2 = 1$.

A hollow waveguide, of rectangular cross section, of height a along Ox, of width b along Oy, of infinite length along Oz is considered (Figure 3.2). The walls of this waveguide are made of a perfect conductor of negligible thickness.

The objective is to study whether a wave expressed by:

$$\vec{E}(M,t) = \vec{E}_0(y) e^{i(\omega t - kz)}, \text{ with } \vec{E}_0(y) = E_0(y) \vec{u}_x, \text{ where } E_0(y) \text{ is real.}$$

and

$$\vec{B}(M,t) = \vec{B}_0(y) e^{i(\omega t - kz)}, \text{ with } \vec{B}_0(y) = B_{0y}(y) \vec{u}_y + B_{0z}(y) \vec{u}_z$$

can propagate in this waveguide.

Consider $k_c^2 = \dfrac{\omega^2}{c^2} - k^2$, where k_c is real and positive.

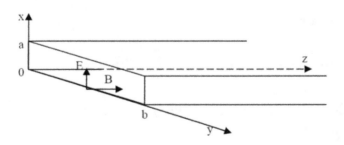

Figure 3.2. *Rectangular waveguide*

1) Does the form given for the wave verify Maxwell equations?

2) Based on the equation of propagation of the electric field E, find the differential equation verified by $E_0(y)$.

Electromagnetic Waves and their Applications 105

3) Using the boundary conditions of the electric field E in $y = 0$ and $y = b$, prove that k_c is an integer multiple of π/b.

4) Deduce the expression of $E_0(y)$.

5) Using **rot E**, determine the components of $B_0(y)$.

6) Express k as a function of ω, c, b and n (n is an integer).

7) Deduce the existence of a cutoff frequency f_c below which propagation is no longer possible.

8) Calculate f_c for $b = 2$ cm.

9) Determine the wave phase velocity as a function of ω, c, n and b.

10) Determine the wave group velocity as a function of ω, c, n and b.

11) What is the relation between phase velocity and group velocity?

12) Calculate the average value in time of the Poynting vector at a point M of the waveguide.

13) Deduce the average power (P) transmitted by a section of the waveguide.

14) Calculate the time average electromagnetic energy (W) in the waveguide per unit length.

15) Define in this case the speed of energy propagation and prove that it is simply expressed as a function of $<P>$ and $<W>$. Comment.

3.4.1. *Answer to question 1*

There are four Maxwell equations:

$$div\ \vec{E} = \frac{\rho}{\varepsilon_0}, \quad div\ \vec{B} = 0, \quad \vec{rot}\ \vec{E} = \frac{-\partial \vec{B}}{\partial t} \quad \text{and} \quad \vec{rot}\ \vec{B} = \left(\mu_0\ \vec{j} \right) + \mu_0 \varepsilon_0 \frac{\partial \vec{E}}{\partial t}$$

The electric field $\vec{E}$ has only one component directed along $\vec{x}$, which leads, after the application of the div operator, to the following result:

$$div(\vec{E}(M,t)) = \frac{\partial}{\partial x} E_0 \left(y \right) e^{i(\omega t - kz)} = 0$$

This means that charges are necessarily absent. This condition is verified, since in the waveguide, there are no free charges and the walls being metallic, there are no charges in the volume of a perfect metal, of infinite conductivity.

For the magnetic induction $\vec{B}$ field, there are two components along $\vec{y}$ and $\vec{z}$, which leads after the application of the operator div to the following result:

$$div(\vec{B}(M,t)) = e^{i(\omega t - kz)} \frac{\partial}{\partial y} E_0(y) + E_0(y) \frac{\partial}{\partial z} e^{i(\omega t - kz)}$$

$$= \left(\frac{dE_0(y)}{dy} + ikE_0(y) \right) e^{i(\omega t - kz)}$$

This relation is not possible unless $div\ \vec{B} = 0$, which means $\left(\dfrac{dE_0(y)}{dy} + ikE_0(y) \right) = 0$.

According to Maxwell's third equation:

$$\overrightarrow{rot}\ \vec{E} = \begin{pmatrix} \frac{\partial}{\partial x} \\ \frac{\partial}{\partial y} \\ \frac{\partial}{\partial z} \end{pmatrix} \wedge \begin{pmatrix} E_x \\ 0 \\ 0 \end{pmatrix} = \begin{pmatrix} 0 \\ \frac{\partial}{\partial z} E_x \\ -\frac{\partial}{\partial y} E_x \end{pmatrix} = \begin{pmatrix} 0 \\ -ikE_0(y)e^{i(\omega t - kz)} \\ -\frac{\partial E_0(y)}{\partial y} e^{i(\omega t - kz)} \end{pmatrix}$$

$$= i\omega \begin{pmatrix} 0 \\ B_{0y}(y)e^{i(\omega t - kz)} \\ B_{0z}(y)e^{i(\omega t - kz)} \end{pmatrix}$$

This means that the amplitudes of the components in y and z of the magnetic induction field depend on those of field E or on its derivative with respect to y.

According to Maxwell's fourth equation:

$$\overset{\rightarrow}{rot}\ \vec{B} = \begin{pmatrix} \partial/\partial x \\ \partial/\partial y \\ \partial/\partial z \end{pmatrix} \wedge \begin{pmatrix} 0 \\ B_y \\ B_z \end{pmatrix} = \begin{pmatrix} \dfrac{\partial B_z}{\partial y} - \dfrac{\partial B_y}{\partial z} \\ 0 \\ 0 \end{pmatrix} = \begin{pmatrix} \dfrac{\partial B_{0z}(y)}{\partial y} e^{i(\omega t - kz)} + ikB_{0y}(y) e^{i(\omega t - kz)} \\ 0 \\ 0 \end{pmatrix}$$

$$= \mu_0 \vec{j} - i\omega\mu_0\varepsilon_0 \begin{pmatrix} E_0(y) e^{i(\omega t - kz)} \\ 0 \\ 0 \end{pmatrix}$$

This means that the current j is necessarily directed along Ox and the amplitudes of the component in y and of the derivative with respect to y of the component in z of the magnetic induction field depend on those of the field E and on the amplitude of the current.

There is no current in a vacuum, with the possible exception of the walls.

3.4.2. *Answer to question 2*

The expression of the equation of propagation of the electric field $\vec{E}$ is:

$$\Delta \vec{E} - \frac{\partial}{\partial t}\left(\frac{1}{c^2} \frac{\partial \vec{E}}{\partial t} \right) = 0$$

The electric field $\vec{E}$ being linearly polarized along $\vec{x}$, only one equation remains, and its form, being expanded with respect to component x, is given by:

$$\left(\frac{\partial^2}{\partial x^2} + \frac{\partial^2}{\partial y^2} + \frac{\partial^2}{\partial z^2} - \frac{1}{c^2} \frac{\partial^2}{\partial t^2} \right) E_0(y) e^{i(\omega t - kz)} = 0$$

108 Applications and Metrology at Nanometer Scale 1

Consequently, after calculating the derivatives, the following equation is obtained:

$$\left(\frac{\partial^2}{\partial y^2}E_0(y)-k^2E_0(y)+\frac{\omega^2}{c^2}E_0(y)\right)e^{i(\omega t-kz)}=0$$

which is the equation verified by the amplitude $E_0(y)$ expressed as:

$$\frac{\partial^2}{\partial y^2}E_0(y)+k_c^2E_0(y)=0$$

where: $k_c^2=\dfrac{\omega^2}{c^2}-k^2$.

3.4.3. Answer to question 3

The boundary conditions correspond to the conservation of the tangential component of $\vec{E}$, on the metal walls. Since $\vec{E}=\vec{0}$ in a metal (considered as perfect, of infinite conductivity), then $\vec{E}=\vec{0}$ on the walls in $x=0$ and $x=a$ and on the walls in $y=0$ and $y=b$. This condition applies only to the tangential component of $\vec{E}$. Since the electric field $\vec{E}$ is tangent to the wall in y, at the interface between the two media, metal and air, the following relation is obtained:

$$\vec{E}_{t(air)}=\vec{E}_{t(metal)}$$

Hence: $\vec{E}(y=0)=\vec{0}$ and $\vec{E}(y=b)=\vec{0}$.

3.4.4. Answer to question 4

The objective is to solve the differential equation:

$$\frac{\partial^2}{\partial y^2}E_0(y)+k_c^2E_0(y)=0 \qquad \text{with} \qquad \text{the} \qquad \text{boundary} \qquad \text{conditions:}$$

$E_0(b)=E_0(0)=0$.

Electromagnetic Waves and their Applications   109

The solution having the form e^{ry}, the differential equation leads to the characteristic equation: $\left(r^2 + k_c^2 \right) = 0$, which yields the complex solution:

$$E_0(y) = Ae^{jk_c y} + Be^{-jk_c y},$$

where A and B are complex conjugates or in cosine form, as: $E_0(y) = C\cos(k_c y + \phi)$, where C and φ are related to the coefficients A and B of the complex solution.

Since: $E_0(0) = 0$, then: $C\cos(0 + \phi) = 0$.

which yields $\phi = \pm \dfrac{\pi}{2}$, leading to $E_0(y) = C\sin(k_c y)$.

The second boundary condition $E_0(b) = 0$ requires $E_0(b) = C.\sin(k_c b) = 0$.

If $C \neq 0$, then: $k_c b = p\pi$ and p being an integer.

The boundary condition $y = b$ leads to:

$k_c = \dfrac{p\pi}{b}$, which leads to the following expression for the amplitude of the field:

$$E_0(y) = A.\sin\left(\frac{p\pi}{b} y \right)$$

3.4.5. Answer to question 5

The expression of $\vec{E}(y, z) = C\sin\left(\dfrac{p\pi}{b} y \right) e^{i(\omega t - kz)} \vec{u}_x$ and the relation

$\vec{rot}\, \vec{E} = \dfrac{-\partial \vec{B}}{\partial t}$ lead to:

$$\overrightarrow{rot}\ \overrightarrow{E} = \begin{pmatrix} \partial/\partial x \\ \partial/\partial y \\ \partial/\partial z \end{pmatrix} \wedge \begin{pmatrix} E_x \\ 0 \\ 0 \end{pmatrix} = \begin{pmatrix} 0 \\ \partial/\partial z\, E_x \\ -\partial/\partial y\, E_x \end{pmatrix} = \begin{pmatrix} 0 \\ -ikC\sin\left(\dfrac{p\pi}{b}y\right)e^{i(\omega t - kz)} \\ \dfrac{-p\pi}{b}C\cos\left(\dfrac{p\pi}{b}y\right)e^{i(\omega t - kz)} \end{pmatrix}$$

$$\overrightarrow{rot}\ \overrightarrow{E} = \frac{-\partial \overrightarrow{B}}{\partial t} = -i\omega \overrightarrow{B}$$

This leads to: $\overrightarrow{B} = \dfrac{-1}{i\omega}.\overrightarrow{rot}\ \overrightarrow{E}$

$$\text{Hence, the expression: } \overrightarrow{B} = \begin{vmatrix} 0 \\ \dfrac{k}{\omega}C\sin\left(\dfrac{p\pi}{b}y\right)e^{i(\omega t - kz)} \\ \dfrac{-ip\pi}{b\omega}C\cos\left(\dfrac{p\pi}{b}y\right)e^{-i(\omega t - kz)} \end{vmatrix}$$

$$B_{0y}(y) = \frac{k}{\omega}C\sin\left(\frac{p\pi}{b}y\right)$$

and $B_{0z}(y) = \dfrac{-ip\pi}{b\omega}C\cos\left(\dfrac{p\pi}{by}\right)$.

3.4.6. *Answer to question 6*

Since $k^2 = \dfrac{\omega^2}{c^2} - k_c^{\ 2}$, with the angular frequency $\omega = 2\pi f$ and $k_c = {n\pi}/{b}$,

hence $k^2 = \dfrac{\omega^2}{c^2} - \left(\dfrac{n\pi}{b}\right)^2$.

3.4.7. *Answer to question 7*

The relation between the frequency f and the angular frequency is $\omega = 2\pi f$.

Thus: $\dfrac{\omega^2}{c^2} - k_c^2 = \dfrac{\omega^2}{c^2} - \left(\dfrac{n\pi}{b}\right)^2 = k^2 \geq 0$,

If: $\dfrac{\omega^2}{c^2} \geq k_c^2$ then $\dfrac{2\pi f}{c} \geq \dfrac{n\pi}{b}$

$f \geq \dfrac{nc}{2b}$ $f \geq f_C = \dfrac{c}{2b}$ for n=1

Therefore, frequency f must be above $f_C = \dfrac{c}{2b}$.

3.4.8. *Answer to question 8*

$f_C = \dfrac{c}{2b}$;

$f_c = (3 \ 10^{10}/4) = 7.5$ GHz (domain of centimeter waves).

No wave of wavelength above 4 cm can propagate in the waveguide.

3.4.9. *Answer to question 9*

The expression of the phase speed V_ϕ is: $V_\phi = \omega/k$

The expression of the wave vector (question 6) is: $k = \sqrt{\dfrac{\omega^2}{c^2} - \left(\dfrac{n\pi}{b}\right)^2}$

Replacing the expression of the wave vector in the expression of the phase velocity: $V_\phi = \omega/k = \dfrac{\omega}{\sqrt{\dfrac{\omega^2}{c^2} - \left(\dfrac{n\pi}{b}\right)^2}}$

3.4.10. *Answer to question 10*

The expression of the group velocity V_g is: $V_g = d\omega/dk$.

From the expression: $\dfrac{\omega^2}{c^2} = k^2 + \left(\dfrac{n\pi}{b}\right)^2$, it is deduced that: $\dfrac{2\omega}{c^2} \cdot \dfrac{d\omega}{dk} = 2k$

hence: $V_g = d\omega/dk = \dfrac{k}{\omega} c^2 = c^2 \dfrac{\sqrt{\dfrac{\omega^2}{c^2} - \left(\dfrac{n\pi}{b}\right)^2}}{\omega} = \dfrac{c}{\omega}\sqrt{\omega^2 - \dfrac{n^2\pi^2 c^2}{b^2}} = \dfrac{c^2}{V_\phi}$.

3.4.11. *Answer to question 11*

The relation between phase velocity and group velocity is: $V_\phi.V_g = c^2$.

3.4.12. *Answer to question 12*

The Poynting vector is given by: $\vec{R} = \dfrac{1}{\mu_0} \mathrm{Re}\left(\vec{E}\right) \wedge \mathrm{Re}\left(\vec{B}\right)$.

According to equation [4.23], page 80 of *Nanometer-scale Defect Detection Using Polarized Light*:

$$R_e\left(\vec{E}\right) = \begin{pmatrix} A\sin\left(\dfrac{n\pi}{b}y\right)\cos\left(\omega t - kz\right) \\ 0 \\ 0 \end{pmatrix}, A \text{ being a real number.}$$

and:
$$R_e\left(\vec{B}\right) = \begin{pmatrix} 0 \\ A\dfrac{k}{\omega}\sin\left(\dfrac{n\pi}{b}y\right)\cos\left(\omega t - kz\right) \\ -\dfrac{An\pi}{\omega}\cos\left(\dfrac{n\pi}{b}y\right)\sin\left(\omega t - kz\right) \end{pmatrix}, \text{ A being a real number.}$$

Hence:
$$\vec{R} = \frac{1}{\mu_0}\begin{pmatrix} 0 \\ \dfrac{A^2 n\pi}{\omega}\sin\left(\dfrac{n\pi}{b}y\right)\cos\left(\dfrac{n\pi}{b}y\right)\cos\left(\omega t - kz\right).\sin\left(\omega t - kz\right) \\ \dfrac{A^2 k}{\omega}\sin^2\left(\dfrac{n\pi}{b}y\right)\cos^2\left(\omega t - kz\right) \end{pmatrix}$$

3.4.13. *Answer to question 13*

$$\langle P\rangle = \langle\vec{R}\cdot s\vec{u}_z\rangle = \frac{ab}{2}\frac{A^2 k}{\mu_0\omega}\left\langle\sin^2\left(\frac{n\pi}{b}y\right)\right\rangle = \frac{1}{2}\frac{ab}{2}\frac{A^2 k}{\mu_0\omega} \text{ with } s = ab$$

3.4.14. *Answer to question 14*

The expression of the average energy density is:

$$u = \left\langle\frac{1}{2}\varepsilon_0\left(R_e\left(\vec{E}\right)\right)^2 + \frac{1}{2}\left(R_e\left(\vec{B}\right)\right)^2 \times \frac{1}{\mu_0}\right\rangle = \varepsilon_0\left\langle\vec{E}^2\right\rangle = \frac{\varepsilon_0 A^2}{2}\left\langle\sin^2\left(\frac{n\pi y}{b}\right)\right\rangle = \frac{1}{2}\frac{\varepsilon_0 A^2}{2}$$

Multiplying by the volume of the unit length, abx1, the average electromagnetic energy is given by:

$$<W> = <u>\cdot s \times 1 = ab\frac{\varepsilon_0 A^2}{2}\left\langle\sin^2\left(\frac{n\pi y}{b}\right)\right\rangle = \frac{1}{2}ab\frac{\varepsilon_0 A^2}{2}$$

3.4.15. *Answer to question 15*

The speed of energy propagation is given by the ratio between the module of the Poynting vector in the direction perpendicular to the surface crossed and the energy density: $V_e = \left\langle \dfrac{|\vec{R}|}{u} \right\rangle$

As: $\left\langle \vec{R} \cdot \vec{u}_z \right\rangle = \dfrac{1}{2} \dfrac{A^2 k}{\mu_0 \omega} \left\langle \sin^2\left(\dfrac{n\pi}{b} y\right) \right\rangle = \dfrac{1}{2} \dfrac{A^2}{\mu_0 c} \left\langle \sin^2\left(\dfrac{n\pi}{b} y\right) \right\rangle$

and

$$u = \left\langle \dfrac{1}{2}\varepsilon_0 \left(R_e\left(\overrightarrow{E}\right)\right)^2 + \dfrac{1}{2}\left(R_e\left(\overrightarrow{B}\right)\right)^2 \times \dfrac{1}{\mu_0} \right\rangle = \varepsilon_0 \left\langle \overrightarrow{E}^2 \right\rangle = \dfrac{\varepsilon_0 A^2}{2}\left\langle \sin^2\left(\dfrac{n\pi y}{b}\right)\right\rangle$$

$$V_e = \left(\dfrac{1}{2}\dfrac{A^2}{\mu_0 c}\left\langle \sin^2\left(\dfrac{n\pi}{b} y\right)\right\rangle\right) \bigg/ \left(\dfrac{1}{2}\varepsilon_0 A^2 \left\langle \sin^2\left(\dfrac{n\pi}{b} y\right)\right\rangle\right) = \dfrac{1}{\varepsilon_0 \mu_0 c} = c$$

As: $\left\langle P \right\rangle = \left\langle \vec{R} \cdot s\vec{u}_z \right\rangle = \dfrac{ab}{2}\dfrac{A^2 k}{\mu_0 \omega}\left\langle \sin^2\left(\dfrac{n\pi}{b} y\right)\right\rangle$

$$<W> = <u> \cdot s \times 1 = ab\dfrac{\varepsilon_0 A^2}{2}\left\langle \sin^2\left(\dfrac{n\pi y}{b}\right)\right\rangle$$

hence: $V_e = \dfrac{<P>}{<W>} = \dfrac{1}{\mu_0}\cdot\dfrac{k}{\omega}\cdot\dfrac{1}{\varepsilon_0} = \dfrac{1}{c\mu_0\varepsilon_0} = c$.

3.5. Characteristics of microwave antennas

Microwaves correspond to the spectrum of sub-meter waves down to millimeter waves, which is the frequency range between 300 MHz and 300 GHz. Microwave antennas are devices that convert electric energy into electromagnetic energy and are used in a wide frequency range (30 kHz to 300 GHz) in communication systems: radio, television, wireless telephony, geo-localization, satellite observation or communication, radar detection, remote sensing [SIL 49], [MAR 90], [BRA 78]. Microwave antennas are also used for the characterization of matter–electromagnetic radiation interaction and for the detection of nanometer-scale defects.

3.5.1. *Introduction to antennas*

A transmission antenna is a device that converts electromagnetic fields propagating through a waveguide or a line into radiation in the environment surrounding the antenna. The phenomena involved in an antenna are reversible. If an electromagnetic wave front is intercepted by the radiative elements of a receiving antenna, the latter picks up the energy of the incident electromagnetic wave and transfers the corresponding electromagnetic field to the transmission line. In practice, an antenna behaves simultaneously as an emitter and a receiver.

Antennas are composed of radiative elements of various designs or forms: wire, dipole, loop, helix, spiral, monopole, cone (pyramid, circular cone), rectangular waveguide, micro-strip, with parabolic reflector, etc. Depending on their design, antennas can be omnidirectional or directional. Omnidirectional antennas can be fixed. Directional antennas can be motor-driven and oriented towards the emitter or the area to cover. The systems using antennas vary greatly. Depending on the applications, there are passive antennas, active antennas with preamplifier, which are used more often in reception, and antennas arranged in networks or matrices.

The structure of an antenna is composed of conductive elements or resonant cavities. In the emission mode, these elements convert the electric currents at their surface into electromagnetic radiation. Conversely, in reception mode, the electromagnetic waves picked up by the conductive elements or by the resonant structures are converted into an electric signal. At the resonance frequency of the antenna circuit, the electric impedance of the antenna becomes real and therefore purely resistive. The condition for an antenna to be efficient is that its input resistance is adapted to the impedance of the waveguide or of the transmission line of the emitting circuit.

An antenna is often characterized by its polarization direction, its input impedance, its bandwidth, its efficiency, its gain and its directivity or its radiation diagram.

The polarization of a receiving antenna is the direction of the electric field that maximizes reception. The direction of polarization of a transmission antenna is the direction of the radiated electric field. When the radiated electric field is parallel to the ground, the polarization of the antenna is vertical. When the radiated electric field is perpendicular (90°) to the ground, polarization is horizontal. There are antennas with circular polarization.

An antenna often operates in a quite low range of frequencies. The bandwidth of an antenna is represented by the two limits of frequencies between which this antenna can operate without efficiency loss. In this range of frequencies, the impedance of the antenna is real. Beyond this bandwidth, the antenna behaves as a series resonant circuit whose impedance (Z_a) has a real part (R_a) and an imaginary part (X_a). The impedance of the antenna is:

$$Z_a = R_a + iX_a$$

To avoid reflection phenomena, the impedance of an emission antenna must be adapted to the impedance of the environment in which the waves propagate (in a vacuum, the impedance is 377 Ohm).

The impedance of a receiving antenna must be adapted to the impedance of the environment in which the waves propagate and to the impedance of the circuit that amplifies the electric signal picked up by the antenna. If these conditions are met, there is no reflection of electromagnetic waves and the maximum radiated power can be transferred.

If $Z_r = R_r + jX_r$ is the impedance of the receiving circuit and $Z_a = R_a + jX_a$ is the impedance of the receiving antenna, these relations are expressed by: $Z_a = Z_r^*$

$$Z_r^* = R_r - jX_r = R_a + jX_a$$

or:

$$R_r = R_a$$
$$jX_r = -jX_a$$

Let us consider an emission antenna, located at the center O of a spherical reference system (Figure 3.3). Oz axis corresponds to the vertical axis of the reference system. The horizontal plane corresponds to the set of points where $\theta = \dfrac{\pi}{2}$ and ϕ is arbitrary. The sets of points defined by an angle θ between 0 and π, and an angle ϕ constant (respectively, -ϕ), in a spherical reference system, are in a vertical plane.

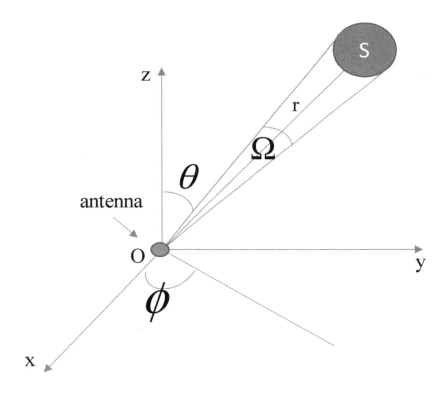

Figure 3.3. *Antenna at the center of a spherical reference system: definition of angles θ and ϕ. The solid angle of the pick-up surface at point M is $\Omega = \dfrac{S}{r^2}$. For a color version of this figure, see www.iste.co.uk/dahoo/metrology1.zip*

The power P_A radiated in a direction defined by the angles (θ, ϕ) can be expressed in a solid angle Ω by the relation:

$$P(\theta, \phi, \Omega) = \frac{P_A}{\Omega}$$

The power density radiated through an elementary surface at a distance r is:

$$p(\theta, \phi, r) = \frac{P_A}{\Omega r^2}$$

The total radiated power P_r is:

$$P_r = \iint_{\theta \, \phi} P(\theta, \phi) \, d\theta \, d\phi$$

For the specific case of an isotropic lossless antenna, the power per unit solid angle is: $P(\theta, \phi) = \frac{P_A}{4\pi}$, where P_A is the electric power supplied, and the power density per unit solid angle $p(\theta, \phi, r)$ at a distance r is:

$$p(\theta, \phi, r) = \frac{P_A}{4\pi r^2}$$

The power received by an antenna is expressed as a function of the effective reception surface $A_{e\!f\!f}$ $\left[m^2 \right]$ and of the intensity of the received power $I(\theta, \phi)$ $\left[W / m^2 \right]$ by the relation:

$$P_r = I(\theta, \phi) A_{e\!f\!f} \left[W \right]$$

The effective reception surface depends on (θ, ϕ) and can be described by $A_{e\!f\!f}(\theta, \phi)$.

In the case of an antenna of length dl, the picked-up voltage V_r is $V_r = E.dl$.

If $R_r = R_a$, the received power P_r is:

$$P_r = \frac{V_{eff}^2}{4R_r} = \frac{E^2 dl^2}{4R_r}$$

The received power is:

$$P_r = A_{eff} \frac{E_{eff}^2}{\sqrt{\dfrac{\mu_0}{\varepsilon_0}}}$$

Since the resistance of radiation in vacuum a is:

$$R_r = \frac{2\pi}{3} Z_0 \frac{dl^2}{\lambda^2}$$

Then:

$$A_{eff} = \frac{dl^2 \lambda^2 \sqrt{\dfrac{\mu_0}{\varepsilon_0}}}{4\left(\dfrac{2\pi}{3} dl^2 \sqrt{\dfrac{\mu_0}{\varepsilon_0}}\right)} = \frac{3\lambda^2}{8\pi}$$

An antenna is often characterized by its gain. The gain G in a direction of solid angle defined by (θ, ϕ) is the ratio between the power P_A radiated per unit solid angle and the power supplied to the antenna divided by 4π.

$$G(\theta, \phi) = \frac{\left(\dfrac{P_A(\theta, \phi)}{\Omega}\right)}{\left(\dfrac{Supplied\,Power}{4\pi}\right)}$$

This definition of gain involves the comparison between the power radiated by a unit solid angle and that of an isotropic antenna of the same power. Examples of an isotropic antenna are an emission–reception WIFI antenna in a residential building or a radar antenna for airplane detection in an airport. The gain of a real antenna is often compared to the gain of an isotropic antenna, such as a dipole or a wire antenna. The gain is then expressed in decibels (dB).

$$G = 10 \log_{10} \frac{P}{P_{ideal}} (dB)$$

If the reference ideal antenna is a dipole, the unit is the dBd (decibel with respect to dipole). In the case of an isotropic reference antenna, it is the dBi (decibel with respect to the isotropic antenna). The difference between dBi and dBd is 2.15 decibels, which means that a dipole whose length is half a wavelength has a gain of 2.15 dBi.

For certain applications, it is interesting to channel the emitted radiation in a privileged direction, so that a high gain emission/detection direction is available. It is, for example, the case of a radar detection antenna. The gain of a directive antenna G_D in a direction (θ, ϕ) is defined by:

$$G_D (\theta, \phi) = \frac{(P(\theta, \phi))}{\left(\dfrac{Total\ radiated\ power}{4\pi} \right)}$$

The gain of a lossless directive antenna whose power is concentrated in a single beam (characterized by a cone of solid angle Ω_s) in which power density is evenly distributed is:

$$G_D = \frac{4\pi}{\Omega_s}$$

A directive antenna has one or two lobes clearly more significant than the others referred to as "main lobes". The narrower the most directive lobe of an antenna is, the more directive is this antenna. The antenna beamwidth characterizes the width of the main lobe. The beamwidth at 3 dB represents the portion of space in which most of the power is radiated. It is the angle between the two directions of the main lobe where the radiated power is equal to half the power radiated in the direction of the maximal radiation. Directivity corresponds to the width of the main lobe principal, between the attenuation angles at 3 dB.

The area (S) illuminated by the beam emitted by an ideal antenna is the product of the solid angle (Ω_R) and the distance (r) between the antenna and the radiated area to the square.

The area illuminated (S) by the beam emitted by an ideal antenna is the product of the solid angle (Ω_R) and the squared distance (r) between the antenna and the radiated area:

$$S = r^2 \Omega_R$$

If the beam emitted by the antenna has a circular section, then the width of the beam θ_R is given by the relation:

$$\Omega_R = \frac{\pi}{4} \theta_R{}^2$$

The radiation diagram of an antenna is a characteristic function of the radiation depending on angles (θ, ϕ), which varies along the direction between 0 and 1. A three-dimensional display of antenna lobes is thus possible, in the horizontal plane or in the vertical plane including the most significant lobe. The power radiation diagram represents the distribution of power per unit solid angle in the direction of the solid angle. The radiation diagram is determined based on the power emitted or received per unit solid angle in the direction of the solid angle.

An antenna is characterized by its near field diagram (in the proximity of the antenna, in the zone where the electric field varies inversely with distance) and by its far field diagram (far from the antenna in the zone where the electric field varies inversely with distance to the square).

The diagram of the near field radiated in the proximity of the antenna can be used to evaluate the impedance of the antenna and its reactive power.

A microwave transmission antenna converts the waves propagated in a line into waves radiated in the environment. Experimentally, a radiation is generated by a time variable electric current or by a charge oscillation.

By definition, the density of electric current in a conductor is defined by an electric charge density ρ_s traveling with a speed V through a section (S) of the conductor:

$$J_S = \rho_s V \ \left(A/m^2\right)$$

In a perfect electric wire, the electric charge density ρ_s can be assumed to be constant. The current in the wire I_s is then expressed as a function of free charges ρ through the wire and the speed V of these charges by the relation:

$$I_s = J_S.S = \left(\rho_s S\right)V = \rho V \ (A)$$

If the current varies in time, it can be written as:

$$\frac{dJ_s}{dt} = \rho \ \frac{dV}{dt} \left(A/m^2/s\right)$$

According to this equation, an electromagnetic radiation can be created by the variation in time of an electric current or by the variation of the acceleration of electric charges. In summary, if a charge does not move, there is no radiation. If a charge moves at constant speed, there is no radiation if the wire is straight and infinitely long, but if the wire has discontinuities or curves, folds, radiation is then possible. Antenna design uses charge oscillations for transient effects and pulses and variable currents for harmonic variations in time.

3.5.2. *Radiation of a wire antenna*

A current of periodic amplitude $I = I_0 e^{j\omega t} \ (A)$ travels through a wire antenna of length (l):

1) Calculate the magnetic potential generated at a point at a distance r from the antenna (r>> l).

2) Deduce the radiated magnetic and electric fields.

3) Calculate the radiated far electric field. Plot the diagram of the radiated electric field. Calculate the ratio of the modules of the electric field and the radiated magnetic field.

Electromagnetic Waves and their Applications   123

4) Calculate the average radiated power. Plot the diagram of the power radiated by the antenna in far field depending on the angle θ in the vertical plane xoz. Find the beamwidth of the antenna at 3 dB.

5) If the antenna has a resistance R, calculate the power dissipated in the antenna and deduce the antenna resistance (which for a current I through the antenna would dissipate a power equivalent to the radiated power).

6) Calculate the antenna efficiency, its gain G and its directivity G_D.

7) Calculate the electric and magnetic fields radiated in the near field. Calculate the Poynting vector and determine the corresponding radiated power in the near field. Find the equivalent Thévenin impedance $Z_A(\omega)$ of the wire antenna.

3.5.2.1. Answer to question 1

For symmetry reasons, the origin O of the Cartesian reference system is chosen at the bottom of the antenna wire and the Oz axis of the reference system is directed along the antenna wire.

The vector potential at time t and a point M at a distance r from the antenna is generated by a current in the antenna at a previous time t'.

The difference $(t - t')$ is the duration required for the electromagnetic wave to propagate from the antenna to the point M. $(t - t')$ is a delay equal to the distance between the antenna and the point M divided by the speed of light, such that:

$$(t - t') = \frac{r}{c}$$

If $\vec{j}$ is the current density in the antenna wire, the vector potential $\vec{A}$ at the point M is obtained from the expression of the retarded potential integrated over the entire volume (V) of the wire antenna:

$$\vec{A}(r,t) = \frac{\mu_0}{4\pi} \int_{\infty} \frac{\vec{j}\left(\vec{r}', t - \frac{|\vec{r} - \vec{r}'|}{c}\right)}{|\vec{r} - \vec{r}'|} d^3\vec{r}'$$

The point M being distant (far field hypothesis with the distance r >> λ the wavelength), the difference in distance being $\left| \vec{r} - \vec{r}' \right| \approx r$, the expression of A is therefore:

$$\vec{A}(r,t) = \frac{\mu_0}{4\pi r} \int_V \vec{j} \left(t - \frac{r}{c} \right) dV$$

If the antenna current flows through a, the element of cross-section S, then the element of volume current $\vec{j}dV$ is expressed by the relation:

$$\vec{j}dV = (\vec{j}.d\vec{S}).d\vec{l}$$

$$\int_V \vec{j}dV = \int_S \vec{j}.d\vec{S} \int_l d\vec{l} = I\vec{l}$$

Since the current I in the antenna is periodic, it can be expressed in complex notation by (see § 4.2 of Chapter 4 [DAH 16]): $I = I_0 e^{j\omega t}$

The vector potential at the point M is: $\vec{A}(r,t) = \frac{\mu_0}{4\pi} \frac{\vec{l}}{r} I_0 e^{j\omega\left(t - \frac{r}{c}\right)}$

Since the wave vector k of electromagnetic waves is related to the angular frequency ω by the relation $k = \frac{\omega}{c}$, the vector potential in the Cartesian reference system $(\hat{x}, \hat{y}, \hat{z})$ is expressed by:

$$\vec{A}(r,t) = \frac{\mu_0}{4\pi} I_0 l e^{j\omega t} \frac{e^{-jkr}}{r} \hat{z}$$

To simplify the expressions of the radiated electric and magnetic fields, the vector potential $\vec{A}(M)$ is expressed in spherical coordinates $(\hat{r}, \hat{\theta}, \hat{\phi})$ (Figure 3.4).

Hence:

$$\vec{A}(r,t) = \frac{\mu_0}{4\pi} I_0 l e^{j\omega t} \frac{e^{-jkr}}{r} \hat{z} = \frac{\mu_0}{4\pi} I_0 l e^{j\omega t} \frac{e^{-jkr}}{r} (cos\,\theta\,\hat{r} - sin\,\theta\,\hat{\theta})$$

$\vec{A}$ has two components A_r and A_θ that are directed along $\hat{r}$ and $\hat{\theta}$.

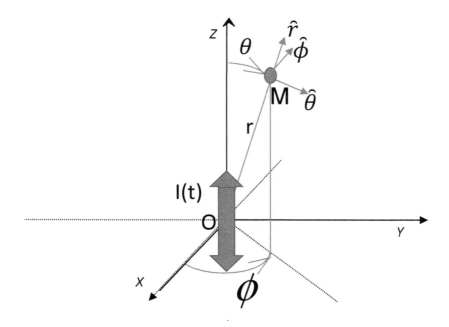

Figure 3.4. *Wire antenna. For a color version of this figure, see www.iste.co.uk/dahoo/metrology1.zip*

3.5.2.2. *Answer to question 2*

The magnetic field is obtained by the relation:

$$\vec{B}(\vec{r},t) = \overrightarrow{rot}\vec{A}(\vec{r},t)$$

In spherical coordinates, this gives:

$$\overrightarrow{rot}\vec{A}(\vec{r},t) = \begin{vmatrix} \dfrac{\hat{r}}{r^2 \sin\theta} & \dfrac{\hat{\theta}}{r \sin\theta} & \dfrac{\hat{\phi}}{r} \\ \dfrac{\partial}{\partial r} & \dfrac{\partial}{\partial \theta} & \dfrac{\partial}{\partial \phi} \\ A_r & rA_\theta & r\sin\theta A_\phi \end{vmatrix}$$

126   Applications and Metrology at Nanometer Scale 1

Since $\vec{A}$ has no component along $\hat{\phi}$ and is independent of position ϕ :

$$A_\phi = 0 \text{ and } \frac{\partial}{\partial \phi} = 0$$

$$rot\vec{A}(r,t) = \frac{\mu_0}{4\pi r} I_0 l \left(\frac{1}{r} \left(\frac{\partial}{\partial r} \left(e^{j(\omega t - kr)} \right) \right) (-\sin\theta) - \frac{\partial}{\partial \theta} \left(\frac{1}{r} e^{j(\omega t - kr)} \right) (\cos\theta) \right) \hat{\phi}$$

$$\vec{B}(r,t) = rot\vec{A}(r,t) = -\frac{\mu_0}{4\pi} I_0 l \sin\theta \left(\frac{1}{r^2} + \frac{jk}{r} \right) e^{j(\omega t - kr)} \hat{\phi}$$

The magnetic field $\vec{H}(r,t)$ is therefore written as:

$$\vec{H}(r,t) = \frac{\vec{B}(r,t)}{\mu_0} = -\frac{I_0 l \sin\theta}{4\pi} \left(\frac{1}{r^2} + \frac{jk}{r} \right) e^{j(\omega t - kr)} \hat{\phi}$$

The magnetic field $\vec{H}(r,t)$ is directed along the direction $\hat{\phi}$ and is therefore perpendicular to the direction of propagation of the radiated wave.

The radiated magnetic field is equal to the real part of $\vec{H}(r,t)$:

$$Re\left(\vec{H}(r,t)\right) = -Re\left(\frac{I_0 l \sin\theta}{4\pi} \left(\frac{jk}{r} \right) e^{j(\omega t - kr)} \hat{\phi} \right) = -\frac{I_0 lk \sin\theta}{4\pi r} \sin(\omega t - kr)\hat{\phi}$$

At the point M, there is no current. The radiated electric field in M is obtained by Ampère's law:

$$\frac{1}{c^2} \frac{\partial}{\partial t} \left(\vec{E}(r,t) \right) = rot\vec{B}(r,t)$$

The only possible dependence of $\vec{E}(r,t)$ with respect to time is from the factor $e^{j\omega t}$ of the current oscillation such that:

$$\vec{E}(r,t) = \vec{E}(r)e^{j\omega t}$$

Electromagnetic Waves and their Applications   127

The magnetic induction radiated at the point M is such that:

$$rot\vec{B}(r,t) = \frac{1}{c^2}\frac{\partial}{\partial t}\left(\vec{E}(r,t)e^{j\omega t}\right) = j\frac{\omega}{c^2}\vec{E} = j\frac{k}{c}\vec{E}$$

This leads to:

$$jk\vec{E} = c\frac{\mu_0}{4\pi}I_0l\left(\frac{1}{r\sin\theta}\frac{\partial}{\partial\theta}\left(\left(\frac{jk}{r}+\frac{1}{r^2}\right)\sin^2\theta\ e^{j(\omega t-kr)}\right)\hat{r}\right.$$
$$\left.+\left(\frac{1}{r}\left(-\frac{\partial}{\partial r}\left(r\left(\frac{jk}{r}+\frac{1}{r^2}\right)\sin\theta\ e^{-jkr}\right)e^{j\omega t}\right)\hat{\theta}\right)\right.$$

$$jk\vec{E} = c\frac{\mu_0}{4\pi}I_0l\left(\left(\frac{1}{r}\right)\left(\frac{jk}{r}+\frac{1}{r^2}\right)(2\cos\theta)e^{j(\omega t-kr)}\hat{r}\right.$$
$$+\left(\frac{1}{r}\left(\left(\frac{1}{r^2}\sin\theta\ e^{-j(\omega t-kr)}\right)\right.\right.$$
$$\left.\left.+ (jk\sin\theta)\left(jk+\frac{1}{r}\right)e^{j(\omega t-kr)}\right)\right)\hat{\theta}\right)$$

$$jk\vec{E} = c\frac{\mu_0}{4\pi}I_0l\left(\left(\frac{jk}{r^2}+\frac{1}{r^3}\right)(2\cos\theta)e^{j(\omega t-kr)}\hat{r}\right.$$
$$+\left(\left(\left(\frac{1}{r^3}\sin\theta\ e^{-j(\omega t-kr)}\right)\right.\right.$$
$$\left.\left.+ (jk\sin\theta)\left(j\frac{k}{r}+\frac{1}{r^2}\right)e^{j(\omega t-kr)}\right)\right)\hat{\theta}\right)$$

$$\vec{E} = -jc\frac{\mu_0}{4\pi k}I_0l\left(\left(\frac{1}{r^3}+\frac{jk}{r^2}\right)(2\cos\theta)e^{j(\omega t-kr)}\hat{r}\right.$$
$$\left.+\left(\sin\theta\left(\frac{1}{r^3}-\frac{k^2}{r}+j\frac{k}{r^2}\right)e^{-j(\omega t-kr)}\right)\hat{\theta}\right)$$

Since the point M is far from the antenna, the distance kr is such that $kr \gg 1$. If λ is the wavelength of the radiation, this means that the point M is at a distance of several wavelengths from the antenna (or $r \gg \dfrac{\lambda}{2\pi}$) and the expressions in $\dfrac{1}{r^2}$ and $\dfrac{1}{r^3}$ of the radiated electric field can be neglected. The electric field can then be written as:

$$\vec{E}(r,t) = jc\frac{\mu_0}{4\pi}I_0 l \sin\theta \frac{k}{r} e^{-j(\omega t - kr)}\widehat{\theta}$$

The radiated electric field is the real part of $\vec{E}(r,t)$:

$$\mathrm{Re}\left(\vec{E}(r,t)\right) = \mathrm{Re}\left(jc\frac{\mu_0 k}{4\pi r}I_0 l \sin\theta e^{j(\omega t - kr)}\right)\widehat{\theta} = -c\frac{\mu_0 k}{4\pi r}I_0 l \sin\theta \sin(\omega t - kr)\widehat{\theta}$$

Therefore, the expressions of the radiated electric and magnetic fields are given by:

$$\vec{H}(r,t) = -\frac{k}{4\pi r}I_0 l \sin\theta \sin(\omega t - kr)\hat{\phi}$$

$$\vec{E}(r,t) = -\sqrt{\frac{\mu_0}{\varepsilon_0}}\frac{k}{4\pi r}I_0 l \sin\theta \sin(\omega t - kr)\hat{\theta}$$

The radiated electric and magnetic fields are in phase and transverse to the direction of propagation. The electric field is directed along $\hat{\theta}$. The magnetic field is directed along $\hat{\phi}$, which is orthogonal to $\hat{\theta}$. Polarization is thus rectilinear. Radiation is symmetrical with respect to ϕ but not with respect to θ. For the far field, radial components are negligible.

3.5.2.3. *Answer to question 3*

The module of the radiated electric field varies with $\sin\theta$:

$$\vec{E}(r) = \sqrt{\frac{\mu_0}{\varepsilon_0}}\frac{k}{4\pi r}I_0 l \sin\theta = E_r \sin\theta$$

The diagram of the radiated far field is obtained by plotting the function $f(\theta) = E_r \sin\theta$. In polar coordinates, this curve is a circle (Figure 3.5). The radiated field is maximum for $\theta = 90°$ and zero along the axis of the antenna.

The ratio of the module of the radiated electric field to the module of the radiated magnetic field $\dfrac{|\vec{E}(r,t)|}{|\vec{H}(r,t)|}$ is equal to the impedance of the medium in which the wave propagates. In a vacuum, this impedance is given by: $Z_0 = \sqrt{\dfrac{\mu_0}{\varepsilon_0}}$ or 377 Ohm.

The electromagnetic radiation has the orthogonality and proportionality properties $\dfrac{|\vec{E}(r,t)|}{|\vec{H}(r,t)|}$ of a uniform plane wave. Nevertheless, the amplitude of the magnetic and the electric far fields decreases with $\dfrac{1}{r}$, which is not the case for a plane wave, and is thus an approximation of the wave front surface by the plane tangent to the point considered.

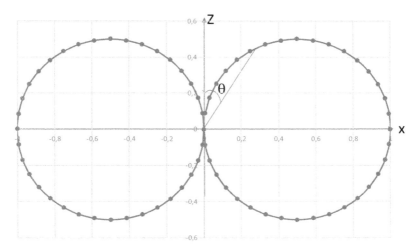

Figure 3.5. *Electric field strength radiated by the wire antenna in the plane defined by the vertical axis z and the horizontal axis x. For a color version of this figure, see www.iste.co.uk/dahoo/metrology1.zip*

3.5.2.4. *Answer to question 4*

The radiated electromagnetic field being a plane wave, the Poynting vector $\vec{S}(\vec{r},t)$ (energy flux per unit surface and time) is expressed as:

$$\vec{S}(\vec{r},t) = \left(\vec{E}(\vec{r},t) \times \vec{H}^*(\vec{r},t)\right) = Z_0 \left(\frac{klI_0}{4\pi r}\right)^2 \sin^2\theta \sin^2(\omega t - kr)\hat{r}$$

The temporal average value of the radiated flux is written as:

$$<\vec{S}(t)> = 0.5\, Re\left(\vec{S}(\vec{r},t)\right) = 0.5\, Re\left(\vec{E}(\vec{r},t) \times \vec{H}^*(\vec{r},t)\right)$$

$$<\vec{S}(t)> = Z_0 \left(\frac{klI_{eff}}{4\pi r}\right)^2 \sin^2\theta \;\; \hat{r} \;\; [W/m^2]$$

with $I_0^2 = 2I_{eff}^2$.

I_{eff} is the effective current in the antenna wire.

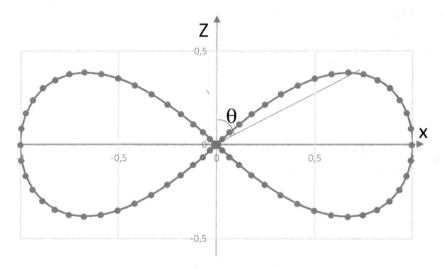

Figure 3.6. *Power radiated by the wire antenna in the plane defined by the vertical axis Z and the horizontal axis X. For a color version of this figure, see www.iste.co.uk/dahoo/metrology1.zip*

The radiated power varies as a function of $\left(\dfrac{kIl_{eff}}{4\pi r}\right)^2 \sin^2 \theta$. The diagram of the power radiated by the antenna in far field is represented in Figure 3.6 by a curve of revolution about axis $\hat{z}$.

If the energy flux is integrated over the entire surface of the sphere of radius r, the power radiated per second is given by:

$$P_r = Z_0 \oint < S(r,\theta) > r^2 d\Omega$$

$$d\Omega = \sin\theta d\theta d\phi$$

$$P_r = \int_0^{2\pi} d\phi \int_0^{\pi} < S(t,r,\theta) > \sin\theta d\theta = \pi Z_0 \left(\frac{kI_{eff}l}{4\pi}\right)^2 \int_0^{\pi} \sin^3\theta d\theta$$

$$P_r = \int_0^{2\pi} d\phi \int_0^{\pi} < S(t,r,\theta) > \sin\theta \, d\theta$$

$$= \pi Z_0 \left(\frac{kI_{eff}l}{4\pi}\right)^2 \int_0^{\pi} \sin^3\theta \, d\theta$$

$$P_r = \pi Z_0 \left(\frac{kIl}{4\pi}\right)^2 \left(\frac{2}{3}\right) = \frac{2\pi}{3} Z_0 \left(\frac{kIl}{4\pi}\right)^2 = \frac{2\pi}{3} Z_0 \left(\frac{2\pi Il}{4\pi\lambda}\right)^2$$

$$= \frac{2\pi}{3} Z_0 \left(\frac{Il}{2\lambda}\right)^2$$

$$P_r = \frac{2\pi\mu_0 c l^2 I_0^2}{3\lambda^2} \frac{1}{2} = \frac{2\pi}{3} \sqrt{\frac{\mu_0}{\varepsilon_0}} \frac{I_0^2 l^2}{2\lambda^2} = \frac{2\pi}{3} \left(Z_0 I_{eff}^2\right) \frac{l^2}{\lambda^2} \, [W]$$

The radiated power (in the far field region) does not depend on distance; radiation extends to infinity. The radiated power is proportional to $\left(\dfrac{I_{eff}l}{\lambda}\right)^2$ and is therefore inversely proportional to the wavelength to the square.

In the horizontal plane, the radiated power is omnidirectional.

In the vertical plane, the beamwidth at 3 dB is the angle θ_{-3dB} for which

$$\frac{P_r(\theta_{-3dB}, \phi)}{P_{max}} = \frac{1}{2} \ 90°.$$

$\theta_{-3dB} = \dfrac{\pi}{2}$. This antenna is not directive.

3.5.2.5. Answer to question 5

If the antenna has a resistance R, the power dissipated in the antenna is:

$$P_R = \left(RI_{eff}^2 \right) [W]$$

The resistance for which the antenna current I would dissipate a power equivalent to the radiated power is known as radiation resistance R_r. This resistance is equal to:

$$R_r = \frac{2\pi}{3} Z_0 \frac{l^2}{\lambda^2} \ [\Omega]$$

3.5.2.6. Answer to question 6

The efficiency η of the antenna is expressed as:

$$\eta = \frac{P_r}{P_r + P_R} = \frac{R_r}{R_r + R_R}$$

The gain G is written as:

$$G(\theta, \phi) = \frac{\left(\dfrac{P_A(\theta, \phi)}{\Omega} \right)}{\left(\dfrac{Supplied\, Power}{4\pi} \right)}$$

The directivity G_D is given by:

$$G_D(\theta,\phi) = \frac{\left(\dfrac{P_A(\theta,\phi)}{\Omega}\right)}{\dfrac{\text{Total radiated power}}{4\pi}}$$

$$G_D = \frac{Z_0\left(\dfrac{2\pi l I_{eff}}{4\pi r \lambda}\right)^2 r^2 \sin^2\theta}{\dfrac{2\pi}{3}\left(Z_0 I_{eff}^2\right)\dfrac{l^2}{\lambda^2}\dfrac{1}{4\pi}} = \frac{3}{2}\sin^2\theta$$

It can be noted that for a wire antenna, the following relation holds:

$$G = G_D t$$

3.5.2.7. *Answer to question 7*

In the near field, the predominant terms vary as $\dfrac{1}{r^3}$

$$\vec{E} \cong -jc\frac{\mu_0}{4\pi k}I_0 l\left(\left(\frac{1}{r^3}\right)(2\cos\theta)e^{j(\omega t - kr)}\hat{r} + \left(\sin\theta\left(\frac{1}{r^3}\right)e^{-j(\omega t - kr)}\right)\hat{\theta}\right)$$

$$\vec{E} \cong -jc^2\frac{I_0 l}{4\pi\varepsilon_0 \omega r^3}\left((2\cos\theta)e^{j(\omega t - kr)}\hat{r} + \left(\sin\theta e^{-j(\omega t - kr)}\right)\hat{\theta}\right)$$

$$\vec{E} = -jc\frac{\mu_0}{4\pi k}I_0 l\left(\begin{array}{l}\left(\dfrac{1}{r^3} + \dfrac{jk}{r^2}\right)(2\cos\theta)e^{j(\omega t - kr)}\hat{r} \\[3mm] + \left(\sin\theta\left(\dfrac{1}{r^3} - \dfrac{k^2}{r} + j\dfrac{k}{r^2}\right)e^{-j(\omega t - kr)}\right)\hat{\theta}\end{array}\right)$$

The dominant term in near field for $\vec{H}(r,t)$ is given by:

$$\vec{H}(r,t) = -\frac{I_0 l \sin\theta}{4\pi}\left(\frac{1}{r^2}\right)e^{j(\omega t - kr)}\hat{\phi}$$

$$\vec{H}(r,t) = -\frac{I_0 l \sin\theta}{4\pi r^2}\hat{\phi}$$

Since the Poynting vector $\vec{S}(\vec{r},t) = \left(\vec{E}(\vec{r},t) \times \vec{H}^*(\vec{r},t)\right)$ is imaginary and negative, near fields correspond to a reactive power and a stored electric energy.

The equivalent Thévenin impedance $Z_A(\omega)$ of the antenna is expressed by:

$$Z_A(\omega) = R_A(\omega) + jX_A(\omega)$$

$R_A(\omega)$ is the resistive part of the impedance corresponding to the total radiated power.

$X_A(\omega)$ is the reactive part corresponding to the energy stored in the near field.

3.6. Characteristics of networks of microwave antennas

The microwave domain covers the range of frequencies between 300 MHz and 300 GHz. The main applications of the networks of microwave antennas are radar systems and the fifth generation (5G) of mobile telecommunication systems.

3.6.1. *Introduction to networks of microwave antennas*

Radar systems use microwave antennas to detect targets. A detection sequence involves microwave radiation emission by the antenna, then reception and analysis of the waves re-emitted by the target.

Radar systems technology went through several development stages. The systems were initially built with parabolic antennas, which were moved mechanically [SIL 49]. Consequently, the use of networks of passive antennas with phase retardation made it possible to control the emission beam direction and to improve the scanning sensitivity of the beam. Due to the network configuration of antennas, emitted waves interfered constructively in the desired target directions and destructively in the complementary space, which lead to an increased directivity of the antenna. The use of drive electronics to control the phase retardation of each antenna of the network made it possible to define the emission/reception beam direction. These systems had no moveable mechanical parts. As the emission and reception electronic modules were often integrated near the antenna network, the maintenance of these radar systems was simplified and their reliability improved. Finally, radar systems were developed around active antenna networks with phase retardation. The presence of an amplifier circuit and a phase adjustment circuit at each antenna of the network enables the numerical control of the emitted or received beam.

Compared to the previous generations (2G, 3G, 4G) of mobile telephony systems, the fifth generation (5G) is a breakthrough technology. 5G telecommunication systems can direct the radiated energy at very high frequency towards the users. They offer very high data transfer rates and very short delays for data transit. They can simultaneously connect to the Internet network a large number of objects. Being compact, they can be installed on the fourth generation equipment towers (Figure 3.7). The frequency bands of the fifth generation (5G) mobile telephony systems are at the present time: 3.3–3.8 GHz and 6–10 GHz.

Figure 3.7. *Block of four 5G antennas operating at 3.5 GHz installed above 4G antennas. For a color version of this figure, see www.iste.co.uk/dahoo/metrology1.zip*

5G antennas are made of networks of antennas in which individual antennas are separated by about one-half wavelength. The addition of a phase retardation to the signal transmitted or received by each antenna enhances the properties of the collective signal of the network of antennas compared to those of individual antennas. The power radiated by the network is higher than that of an isolated antenna. The beam is narrower. It is possible to direct the emission/reception beam at the phase retardation control frequency.

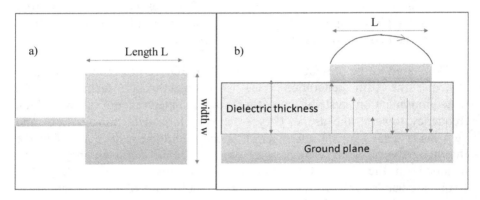

Figure 3.8. *Schematic diagram of a patch antenna: a) top view of the rectangle conductor whose length L is about λ/2 and b) sectional view of the resonant cavity and radiated electric field (blue). For a color version of this figure, see www.iste.co.uk/dahoo/metrology1.zip*

5G antennas are often constituted of networks of patch antennas (Figure 3.8). Patch antennas are planar radiative structures. They can, for example, be obtained by etching on a printed circuit. They are formed of a metallic layer of rectangular, circular or arbitrary geometry, deposited on the dielectric layer of a printed circuit. The opposite face of the printed circuit is metallic and operates as a ground plane. The patch, the ground plane and the four edges form a cavity that can be excited by a microwave of wavelength λ. The length of the patch is of the order of λ/2. The thicknesses of the conducting and dielectric layers are often very small compared to the wavelength. When the patch is supplied by a microstrip line, it enters into resonance, similarly to a half-wave dipole. Excitation generates in phase diffracted electric fields on the edges of the antenna. Due to the ground plane, radiation occurs only in the half-plane above the ground plane. If each antenna has a phase shift circuit and a minimizing circuit, the delay and the amplitude of the current of each antenna in the network can be numerically controlled. The use of numerical control techniques makes it possible to form a beam and move this beam at a very high frequency. Several beams can be emitted simultaneously by a network of antennas. This provides a good signal to noise ratio and improves data transfer rate.

3.6.2. *Radiation of antenna networks*

To illustrate the operating principle of a network antenna with phase retardation, let us consider a network of N equidistant antennas separated by a distance D, supplied by a current of periodic amplitude $I = \frac{I_0}{\sqrt{N}} \cos(\omega t - \varphi_N)$ (see Figure 3.9).

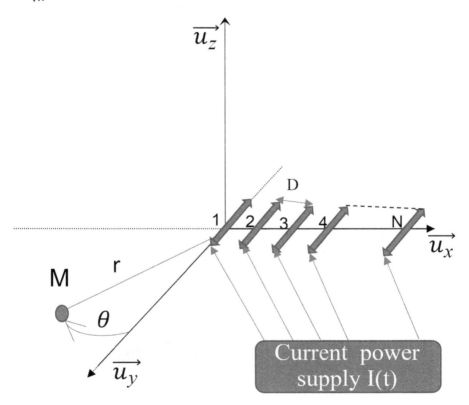

Figure 3.9. *Network of N current-supplied antennas. For a color version of this figure, see www.iste.co.uk/dahoo/metrology1.zip*

The purpose is to study the total electric field radiated at infinity. The antennas are directed along $\vec{u_y}$. They are lined up along $\vec{u_x}$ and the field in plane ($\vec{u_z}, \vec{u_y}$) is studied along a direction given by θ:

1) Assume that $\varphi_N = 0$, for all N. For a point M, at a distance r from the network, with $r \gg ND$, write the delay as a function of I, D,θ c. Calculate E_i^* as a function of E_1^*. Then, deduce the sum of the amplitudes $E(\theta)$. Draw the radiation diagram $|E^*(\theta)|^2$ as a function of θ for a network of eight antennas spaced at a distance $D = \lambda$ for a wavelength λ of 0.03 m in the limit $\left(\frac{D}{\lambda}\right) sin\theta \ll 1$.

2) Delays of $0, \tau, 2\tau, 3\tau, \ldots, (N-1)$ are introduced on the supply line 1,2,3,..,N, respectively (Figure 3.10). What does this mean for the phases φ_N? As $(\varphi_{i-1} - \varphi_i) \leq 2\pi$, deduce the useful range of delays τ. Then, calculate $E_\tau(\theta)$ and $E_\tau^2(\theta)$.

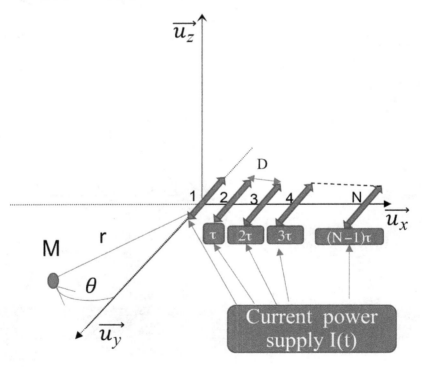

Figure 3.10. *Delays applied to the N antennas of the network. For a color version of this figure, see www.iste.co.uk/dahoo/metrology1.zip*

Draw the new radiation diagram for a delay $\tau = 4.10^{-11}(s)$.

Why is this referred to as beam steering?

3) How does the diagram change if, in addition, the delay system introduces an attenuation a = $(1-\varepsilon)$? Draw the new radiation diagram for a coefficient of attenuation a = $(1-\varepsilon) = 0.95$ and a delay $\tau = 4.10^{-11}(s)$.

4) Draw an analogy of the configuration at question 1) to the light wave falling on an etched plate with each roughness being considered as an individual reflector, at normal incidence. Do the same for the configuration at question 2) at oblique incidence θ_i. Find the relation between θ_i and τ. Finally, consider the case in the limit $D\frac{sin\theta}{l} > 1$.

3.6.2.1. *Answer to question 1*

Each antenna i of the network emits an electric field $\vec{E}_i$ given by:

$$\vec{E}_i = \frac{A}{r_i}\cos\left(\omega(t - \frac{r_i}{c}) + \varphi_i\right)\vec{u_y}$$

For a point M in the plane $(\vec{u_x}, \vec{u_y})$, the amplitude of the electric field $\vec{E}_i$ of the ith antenna of the network is given by:

$$E_i = j\frac{l}{2\lambda}\frac{sin\theta}{r_i}\sqrt{\frac{\mu_0}{v_0}}\frac{I_0}{\sqrt{N}} = \frac{A}{r_i}$$

If the observation point of the radiated field is far from the antenna network, then:

$$\frac{A}{r_i} \approx \frac{A}{r}$$

where r designates the distance from antenna 1 of the network to the point M, and $r \gg ND$.

The electric field radiated by each antenna is: $E_{ref} = j\frac{l}{2\lambda}\frac{sin\theta}{r}\sqrt{\frac{\mu_0}{v_0}}\frac{I_0}{\sqrt{N}}$.

Since vectors $\vec{r_i}$ are parallel (see Figure 3.11):

$$r_2 = r_1 - D\sin\theta$$

By recurrence: $r_i = r_1 - (i-1)D\sin\theta$

and: $\frac{r_i}{c} = \frac{r_1}{c} - (i-1)\frac{D}{c}\sin\theta$

To obtain the contribution $\sum_{i=1}^{N} \vec{E_i}$ of all the antennas in the network, the harmonic expression $\overrightarrow{E_i^*}$ of the electric field parallel to $\vec{u_y}$ radiated by the ith antenna is used.

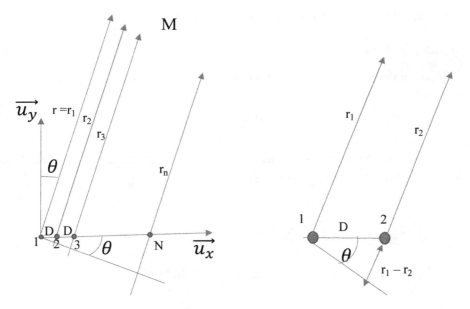

Figure 3.11. *Applied delay. For a color version of this figure, see www.iste.co.uk/dahoo/metrology1.zip*

Let E_i^* be the component y of the radiated electric field:

$$E_i^* = E_{ref} \exp(j\omega\left(t - \frac{r_i}{c}\right) + \varphi_i)$$

As $\varphi_i = 0$,

$$E_i^* = E_{ref}\exp\left(j\omega\left(t - \frac{r_1}{c} - \frac{(i-1)D}{c}\sin\theta\right)\right)$$

$$E_i^* = E_{ref}\exp\left(j\left(\omega(t - \frac{r_1}{c}) - (i-1)\left(\frac{\omega D}{c}\sin\theta\right)\right)\right)$$

Since $\dfrac{\omega D}{c} = \dfrac{2\pi v D}{c} = \dfrac{2\pi D}{\lambda}$

$$E_i^* = E_1^*\exp\left(-2\pi j(i-1)\left(\frac{\omega D}{c}\sin\theta\right)\right)$$

$$= E_1^*\exp\left(-2\pi j(i-1)\left(\frac{D}{\lambda}\right)\sin\theta\right)$$

The field radiated by the network is expressed as:

$$\sum_{i=1}^{N} E_i^* = E_1^*\sum_{i=1}^{N}\exp\left(-2\pi j(i-1)\left(\frac{D}{\lambda}\right)\sin\theta\right)$$

$$= E_1^*\sum_{i'=0}^{N-1}\exp\left(-2\pi j(i')\left(\frac{D}{\lambda}\right)\sin\theta\right)$$

$\sum_{i=1}^{N} E_i^*$ is a geometric series that can be easily summed using the formula: $1+a+\dots+a^{N-1} = \dfrac{1-a^N}{1-a}$

$$\sum_{i=1}^{N} E_i^* =$$

$$E_1^* \frac{1-\exp(-2\pi jN\left(\frac{D}{\lambda}\right)\sin\theta)}{1-\exp(-2\pi j\left(\frac{D}{\lambda}\right)\sin\theta)} =$$

$$E_1^* \frac{\exp\left(-j\pi N\left(\frac{D}{\lambda}\right)\right)}{\exp\left(-j\pi\left(\frac{D}{\lambda}\right)\right)} \frac{\exp(j\pi N\left(\frac{D}{\lambda}\right)\sin\theta)-\exp(-j\pi N\left(\frac{D}{\lambda}\right)\sin\theta)}{\exp(j\pi\left(\frac{D}{\lambda}\right)\sin\theta)-\exp(-j\pi\left(\frac{D}{\lambda}\right)\sin\theta)}$$

$$\sum_{i=1}^{N} E_i^* = E_1^* e^{-\left(j\pi(N-1)\frac{D}{\lambda}\right)} \frac{\sin(\pi N\left(\frac{D}{\lambda}\right)\sin\theta)}{\sin(\pi\left(\frac{D}{\lambda}\right)\sin\theta)}$$

$$E(\theta)^2 = E_{ref}^2 \frac{\sin^2\left(\pi N \left(\frac{D}{\lambda}\right) \sin\theta\right)}{\sin^2\left(\pi \left(\frac{D}{\lambda}\right) \sin\theta\right)}$$

If θ is small and N is large or more precisely if $\left(\frac{D}{\lambda}\right) \sin\theta \ll 1$ or still if $\sin\theta \ll \frac{\lambda}{D}$, then:

$$E(\theta)^2 = E_{ref}^2 \frac{\sin^2\left(\pi N\left(\frac{D}{\lambda}\right)\sin\theta\right)}{\sin^2\left(\pi\left(\frac{D}{\lambda}\right)\sin\theta\right)} = \left(\frac{l}{2\lambda} \frac{\sin\theta}{r}\right)^2 \sqrt{\frac{\mu_0}{\varepsilon_0}} \frac{I_0^2}{N} \frac{\sin^2\left(\pi N\left(\frac{D}{\lambda}\right)\sin\theta\right)}{\left(\pi N\left(\frac{D}{\lambda}\right)\sin\theta\right)^2}$$

$$E(\theta)^2 = \left(\frac{l}{2\lambda} \frac{\sin\theta}{r}\right)^2 \sqrt{\frac{\mu_0}{\varepsilon_0}} \frac{I_0^2}{N} \operatorname{sinc}^2\left(\pi N \left(\frac{D}{\lambda}\right) \sin\theta\right)$$

The first zero is obtained when $N\frac{D}{\lambda} \sin\theta = 1$.

Let us consider: $\sin\theta = \frac{\lambda}{ND}$

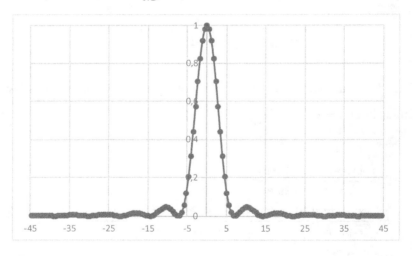

Figure 3.12. *Radiation diagram $E(\theta)^2$ of a network of eight antennas as a function of the angle θ (degrees) for $\frac{D}{\lambda} = 1$. For a color version of this figure, see www.iste.co.uk/dahoo/metrology1.zip*

Figure 3.12 shows the radiation diagram $|E^*(\theta)|^2$ as a function of θ of a network of eight antennas spaced at a distance $D = \lambda$ for a wavelength λ of 0.03 m.

The width of the main peak is: $2\dfrac{\lambda}{ND} \sim \dfrac{2}{N}$ if $\lambda \sim D$.

3.6.2.2. *Answer to question 2*

When there are delays on each antenna of the network, it can be noted that the delay φ_i of the antenna i is expressed by:

$$\varphi_i = \omega\tau_i = \omega(i-1)\tau$$

Such that:

$$\omega(t - \tau_{i)} = \omega t - \omega(i-1)\tau$$

Hence:

$$E_i^* = E_1^* \exp\left(-j(i-1)\frac{2\pi}{\lambda}(D\sin\theta - \tau v)\right)$$

As $v = \dfrac{c}{\lambda}$, τ is significant if $2\pi\tau v \le 2\pi$

$$E_i^* = E_1^* \exp\left(-j(i-1)\frac{2\pi D}{\lambda}\left(\sin\theta - \tau\frac{v}{D}\right)\right), \text{ and:}$$

$$E_i^* = E_1^* \exp\left(-j(i-1)\frac{2\pi D}{\lambda}\left(\sin\theta - \tau\frac{c}{D}\right)\right)$$

$$\sum_{i=1}^{N} E_i^* = E_1^* e^{-(j\pi(N-1)\frac{D}{\lambda})} \frac{\sin(\pi N(\frac{D}{\lambda})(\sin\theta - \tau(\frac{c}{D})))}{\sin(\pi(\frac{D}{\lambda})(\sin\theta - \tau(\frac{c}{D})))}$$

$$E(\theta)^2 = E_{ref}^2 \frac{\sin^2(\pi N\left(\frac{D}{\lambda}\right)\left(\sin\theta - \tau\left(\frac{c}{D}\right)\right))}{\sin^2(\pi\left(\frac{D}{\lambda}\right)(\sin\theta - \tau\left(\frac{c}{D}\right)))}$$

Figure 3.13 represents the radiation diagram of a network of eight antennas spaced at a distance $D = \lambda$ for a wavelength λ of 0.03 m and for a delay $\tau = 4.10^{-11}(s)$. The distribution of the radiated field is shifted from the origin by an angle θ_0 such that $sin\theta_0 = \tau\frac{c}{D}$.

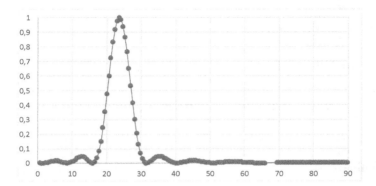

Figure 3.13. Radiation diagram $E(\theta)^2$ of a network of eight antennas as a function of the angle θ (degrees) in the presence of delays τ such that $\tau\frac{c}{D} = 0,4$ for $D = \lambda = 0.03$ (m). For a color version of this figure, see www.iste.co.uk/dahoo/metrology1.zip

The introduction of delays does not change the width of the radiated beam and makes it possible to shift the angular position of the central peak of the radiated field. As the origin of the beam emitted by the network shifts by an angle proportional to the delay, it is possible to control the position of the beam by introducing a progressive phase lag gradient on all the antennas in the network and varying the amplitude of this gradient.

3.6.2.3. Answer to question 3

If an attenuation a(1-ε) is introduced, the amplitude of the emitted field is proportional to:

$$1 + (1-\varepsilon)a + (1-\varepsilon)^2 a^2 + \ldots + (1-\varepsilon)^{N-1} a^{N-1} = \frac{1-(1-\varepsilon)^N a^N}{1-(1-\varepsilon)a}$$

For $\varepsilon = 0.05$, the radiated field is attenuated by a factor of 0.84.

Figure 3.14 represents the radiation diagram of a network of eight antennas spaced at a distance $D = \lambda$ for a wavelength λ of 0.03 m, for a delay $\tau = 4.10^{-11} (s)$ and a coefficient $\varepsilon = 0.05$.

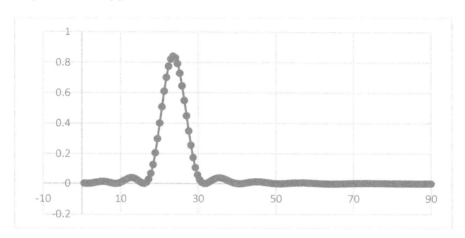

Figure 3.14. *Radiation diagram $E(\theta)^2$ of a network of eight antennas as a function of angle θ (degrees) in the presence of attenuation of 0.05 and delays τ such that $\tau \dfrac{c}{D} = 0,4$ for $D = \lambda = 0.03$ (m). For a color version of this figure, see www.iste.co. uk/dahoo/metrology1.zip*

For a network of antennas in which the amplitude and the phase of each elementary antenna can be changed in real time, it is possible to control the direction and the amplitude of the emitted beam. The use of numerical techniques and coding algorithms makes it possible to simultaneously obtain several beams from the same network of antennas (multiple input multiple output or MIMO technology).

4

Smart Materials

Research work in engineering science on applications of a material for its uses in society focuses on its physical and chemical properties that may be of interest. During the Iron Age, the strength of iron based on its mechanical properties, compared to bronze alloys, was a key criterion in the choice of material for the fabrication of weapons and tools. Later on, the scientific methods developed to process materials in laboratories, particularly in materials science, led to the discovery of semiconductors which paved the way to the electronics industry, and hence, the development of electrical systems based on electronics and computer science. Today, the focus of designers is not only in active materials, but also on their uses in microsystems that make them smart, from which the term smart materials is coined. The main characteristic of these materials is the typical coupling of their various physical properties, which recommends their use as active materials. Piezoelectric materials are best known for their uses. The direct effect was discovered by the Curie brothers in 1880, on the Tourmaline crystal, while the converse effect was theoretically predicted by Lippman based on thermodynamic considerations in 1881, and discovered the same year by the Curie brothers.

4.1. Introduction

Materials are the core of any mechanical, electrical, electronic, optical, magnetic, chemical or thermal components and systems. Materials have always been behind the development level of a civilization. Past history reveals that materials were used for dating the evolution in the human way of life starting with the Stone Age (2 million years), the Copper Age and the

Bronze Age (5000 years), and finally the Iron Age (3000 years). A technological breakthrough occurred around 1850, when a low-cost process was developed for steel manufacturing, leading to the construction of railways and modern infrastructures in the industrial world. The resulting metallurgical industry is still active nowadays, because iron or iron-based alloys are needed in the building, construction or transportation sectors.

Empirical methods were replaced by theoretical and analytical ones based on the experience gained by generations of manufacturers and scientists interested in the development of materials. Throughout the Iron Age, new materials were developed, such as ceramics, semiconductors, polymers or composite materials:

– Metals: valence electrons break free from atoms and delocalize in the conduction band; the ions are maintained in the crystal structure, which is generally of a face-centered cubic (fcc), hexagonal compact (hcp) or cubic-centered (cc) type. Metals are characterized by their mechanical strength and ductility. They have good electrical and heat conductivities. If polished, their surface shines.

– Semiconductors: the bonds are covalent. Electrons are shared by the bonded atoms. Electrical conductivity strongly depends on the proportion of impurities or dopants present in the crystal lattice. Examples are Si, Ge and GaAs. The junction diode opened the way for electronics industry.

– Ceramics: the bonds are ionic between positive ions (cations) and negative ions (anions) due to the Coulomb forces. These materials are composed of metallic cations or semiconductor cations in a lattice constituted of oxygen, nitrogen or carbon anions (oxides, nitrides, carbides). They are hard, brittle and insulating. Examples are glass, porcelain and perovskites.

– Polymers: they are bonded by covalent and van der Waals forces, and they are basically formed of carbon C and hydrogen H atoms; however, they may also contain oxygen O and silicon Si atoms. They break down at moderate temperatures (100–400°C), and they are light. Examples are plastics, rubber and silicone gel.

The development of quantum mechanics, starting in the 1930s, made it possible to understand the behavior of materials and the differences in their properties. Atomic interactions could thus be explained, first in atoms, then in molecules and in solids. The combination of physics and chemistry in the search on the link between the properties of a material and its microstructure is today the field of materials science. This field led to the design of materials and to the knowledge know-how about applications in the field of materials engineering. Industrial processes were developed to manufacture functional devices.

For example, carbon takes various forms: diamond, graphite, nanotubes or fullerenes. The properties of these various forms are related to their underlying structures. The bonds that keep together the atoms in diamond and graphite through shared electrons are covalent.

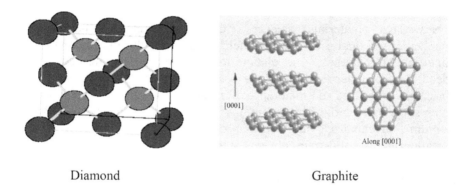

Diamond   Graphite

Figure 4.1. *3D diamond structure of fcc type and 2D graphite structure of plane type, two solid-phase forms of carbon. For a color version of this figure, see www.iste.co.uk/dahoo/metrology1.zip*

In the case of diamond, carbon crystallizes in a face-centered cubic structure (Figure 4.1) and uses its four bond possibilities (sp^3 hybridization), which explains its thermal and mechanical stability. All σ bonds are used so that no delocalized electrons are available for electrical conduction, which is the cause of its insulating nature. In the case of graphite (sp^2 hybridization), carbon uses only three electrons for the σ bonds and structured in the form of planes. The π bond (Figure 4.1) is perpendicular to the planes and provides delocalized electrons responsible for its electrical conductivity. As the planes can slide over one another, graphite is ductile.

From a mechanical point of view, diamond is hard and graphite is ductile, and from an electrical point of view, diamond is insulating, while graphite is conducting. Chapter 9 of [DAH 16] gives an application of carbon nanotubes. Carbon nanotubes can be obtained under high pressure, as indefinitely long wires, which have high mechanical strength.

In order to understand the full complexity of all the naturally occurring materials or synthesized in laboratories, materials are classified into groups. The classification criterion is based on their bonds and not on their structure, properties or uses. There are, however, a few exceptions. For example, for certain materials such as the fiberglass, concrete or wood, the microstructure is a composite of various materials. Biomaterials are any type of biocompatible materials that can replace parts of the human body.

Nowadays, materials science studies the relation between the organization of matter at the nanometric scale and the microstructure and properties of materials. The elaboration of functional materials requires indeed a deep understanding of the relation between structure, properties, fabrication processes and material performance. The materials used in smart systems are characterized by a coupling between various properties. Besides quantum mechanics, the study of material properties involves crystallography, solid-state physics and condensed state matter physics, while the coupling between the properties of the materials is studied in the framework of thermodynamics and statistical physics. Study of the properties of smart materials involves understanding interactions between sub-disciplines in the field of physics and also modeling and simulation.

4.2. Smart systems and materials

A material is qualified as active if under external stimuli it reacts in various ways, according to its nature. When an electric field is applied, its response is a flow of free charges which gives rise to a current if it is a metal or a polarization of matter if it is a dielectric. When a magnetic field is applied, its response is its magnetization if it is a ferromagnetic or ferrimagnetic material. When subjected to a mechanical force, such as the effect of a pressure, it gets more or less deformed if it is a malleable and ductile solid or it breaks if it is rigid. In the case of heat or energy input, its temperature increases either locally, if it is an insulator or throughout its volume and if it is a conductor. Finally, its interaction with an electromagnetic wave leads to absorption, reflection or reflection and refraction. These characteristics can generally be used in technological systems, and the material is qualified as active.

The term smart material refers to an active material, whose properties are embedded in a system involving a sensor, an actuator and an intelligence supplied by an instrumentation and control system (Figure 4.2), which is able to interact with the environment according to the sensed stimuli and adapt to the external stress. Thus, this system is able to interpret or sense physical or chemical variations in its environment and its effects or consequences. If it has built-in intelligence, it must be able to respond to external stress or excitation by means of an active control mechanism. Such a system has no conscience, but in operation, it presents analogies with a biological system in the living world that is capable of appropriate reaction to the changes in its environment. Hence, such a material is qualified as smart.

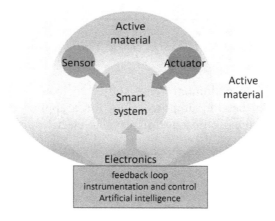

Figure 4.2. *The three elements of a smart system. For a color version of this figure, see www.iste.co.uk/dahoo/metrology1.zip*

However, not all the materials have the functionality that enables them to be integrated in a smart system. The following example illustrates the specific characteristics that a material should have to be qualified as smart material.

Two beams are considered: one made from aluminum (Al) – a metal, and the other one from quartz (SiO_2) – a piezoelectric material. They are both subjected to a mechanical force. The mechanical and electrical responses to this excitation are then studied. The diagrams of the system setup are shown in Figures 4.3 and 4.4.

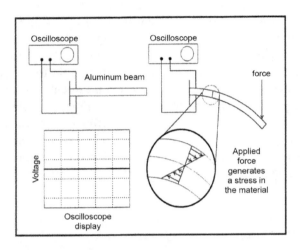

Figure 4.3. *Zero electrical response of an aluminum beam to a mechanical excitation*

In the case of an aluminum beam subjected to a mechanical stress as represented by the bending imposed by a force applied at the end of the beam (Figure 4.3), the oscilloscope displays no electrical response.

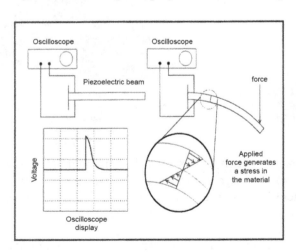

Figure 4.4. *Non-zero response of a piezoelectric beam to mechanical excitation*

In the case of a piezoelectric beam subjected to a mechanical stress as represented by the bending imposed by a force at the end of the beam (Figure 4.4), the oscilloscope displays an electrical response.

The different responses of these two materials – aluminum and piezoelectric material – can be explained, on the one hand, by their structural properties and, on the other hand, by their crystalline symmetry. The structure of aluminum is composed of positive ions or cations linked by a metallic bond to free negative charges, which are peripheral electrons of each aluminum atom, as described in Chapter 5 of [DAH 16]. The applied force induces a mechanical response – namely a bending – as the metal is malleable. Due to the presence of free charges, there is no potential difference at the surface in the absence of an electric field. A piezoelectric material is generally composed of ions that are linked by ionic bonds. Because a mechanical strain is induced by a mechanical force, the material is deformed and as charges are linked by electrostatic forces, the material is polarized. As a result, positive and negative charges are present at the surface. Quartz (SiO_2) is a piezoelectric material. Figure 4.5 shows an example of surface polarization when it is deformed. Because of the mechanical strain, a potential difference can thus be detected – a phenomenon that is observed as an electrical response displayed by an oscilloscope.

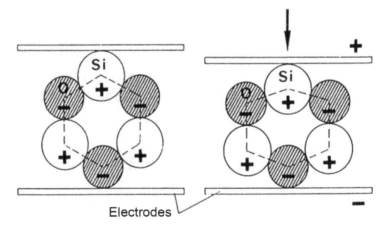

Figure 4.5. *Surface polarization of quartz under mechanical strain*

A metal presents no coupling between its electrical and mechanical properties. In quartz or in a piezoelectric material, there is a coupling between its mechanical and electrical properties. The active materials that are used in smart systems are characterized by the presence of this coupling between two physical or physical and chemical properties of the material. The crystalline structure of the most commonly used piezoelectric crystals is of perovskite type (Figure 4.8) ($CaTiO_3$, $BaTiO_3$, titanates, zirconates, stannates). Piezoelectricity cannot occur in ionic materials, with a high degree of symmetry (this phenomenon requires the absence of a center of symmetry in the unit cell). It is worth noting that the crystalline structure of quartz is hexagonal at high temperature (quartz β) and trigonal at low temperature (quartz α).

The example of quartz shows that for a piezoelectric material, the response is mechanical for an electrical excitation, while an electrical response is obtained for a mechanical excitation. Similar to this electro-mechanical coupling present in quartz, there are other types of couplings, as indicated in Table 4.1: electromagnetic, electro-thermal, electro-optical, thermo-mechanical, pyroelectric, etc. in other materials that can be used as a sensor or an actuator or both.

Following the example of piezoelectric materials, there are (electro-, magneto- or opto-) active polymers, Shape Memory Alloys (SMA), Shape Memory Polymers (SMP), magnetostrictive materials, magneto-rheological fluids, electro-rheological fluids, photochromic materials, polymorphic materials, viscoelastic materials, optical fibers, etc. which are potential candidates to be used as smart materials.

The shape of a piezoelectric or electrostrictive material or of an electroactive polymer changes when voltage is applied. The shape of a shape memory alloy (SMA) changes beyond a certain critical temperature due to a change in structure (phase-transition materials). In magneto-rheological fluids, viscosity changes when a magnetic field is applied. Similarly, the viscosity of electro-rheological fluids changes when an electric field is applied. In a photochromic material, light–matter interaction induces a change in the color of the material. It is worth noting the transition of Micro Electro-Mechanical Systems (MEMS), which contain functional materials towards nanomechanical systems.

Excitation	Response	(Electricity) Current/Charge	(Magnetism) Magnetization	(Mechanics) Strain	(Heat) Temperature	(Optics) Light
(Electricity) Electric field or potential difference		Conductivity Permitivity	Electromagnetic effect	Converse piezoelectric effect Electrostriction Electro-rheological fluid	Electrical resistance Thermoelectric or Peltier effect	Electro-optic Eletcro-chromic Electroluminescent effect
(Magnetism) Magnetic field		Foucault currents	Permeability	Joule effect & Magnetostriction Magneto-rheological fluid	Magnetocaloric effect	Magneto-optic effect
(Mechanics) Stress		Direct piezoelectric effect	Villari effect – reverse magnetostriction	Elasticity modulus Negative Poisson coefficient	Thermomechanical effect	Photoelastic Mechano-chromic effect
(Thermodynamics) Heat		Direct pyroelectric effect Seebeck effect	Thermo-magnetization effect	Thermal expansion and Phase transition	Specific heat	Thermo-luminescence
(Optics) Light		Photovoltaic Photoconductive effect	Photo-magnetization effect	Photostriction	Photo-thermal Thermal Luminescent Thermo-chromic effect	Refraction index Photochromic

Table 4.1. *Smart materials for sensors and actuators*

These materials, which are all characterized by the existence of a coupling between at least two types of properties, have technological applications in many fields, such as electricity and magnetism, electronics, computer science, robotics and mechatronics, technologies related to land and space transportation, and applied science in nanotechnologies (physics, chemistry, biology, etc.). Examples of technological applications are as follows:

– Electrical interconnections, chip carriers, spacer dielectrics, encapsulations, heat dissipaters, heat interface materials, EMI shielding and the casing itself.

– Electric circuit (resistors, capacitors, inductors), electronic devices (diodes, transistors), optoelectronic devices (solar cells, light sensors, electroluminescent diodes) and thermoelectric devices (heaters, coolers, thermocouples).

– Piezoelectric devices (sensors, actuators), micromachines (nanosystems or micro electro-mechanical systems (MEMS)), ferroelectric computer memories, electric interconnections (welding joints, thick film conductors, thin film conductors) and dielectrics (volume, thick film and thin film insulators).

– Substrates for thin films and thick films, heat interface materials in heat dissipaters, cases, electromagnetic interferences (EMI), cables, connectors, electrical supplies, electrical energy storage, motors, electrical contacts (brushes or sliding contacts), etc.

There are various types of actuators as follows: electrostrictive, magnetostrictive, shape memory alloys, viscoelastic materials of magneto-rheological type or electro-rheological type. Sensors can have the form of nanowires in piezoelectric strain gauges or optical fiber strain gauges, constituted of dielectric material structured in the form of digital sensors, in the form of SMA, semiconductor material sensitive to an external physical–chemical stress, in the form of material with giant magneto-impedance (GMI) or giant magneto-resistance (GMR) properties.

There is no single solution for all possible applications. It is sometimes the most appropriate strategy to make an incremental innovation on the existing implementations from initial technologies developed for specific clients and market needs. It is also possible but risky to implement a technology breakthrough generated by fundamental research laboratories. These breakthrough innovations can be applied in various industrial fields in static heavy structures or mobile light structures.

In the field of civil engineering, examples of heavy structures are buildings, bridges, piers, motorways, airport runways and landfill covers. In the field of civil and military industries, examples of mobile light structures used in land, sea, inland waterway, aeronautics or space transportation are as follows: wheelchairs, electric bicycles, automobiles, trains, ships, trucks, tractors, airplanes, submarines, missiles, satellites, transportable bridges and the constituent elements of these constructions such as vehicle body, bumpers, shaft, windows, engine components, brakes and turbine blades. In everyday life, examples of applications include machinery, sport articles, domestic appliances, computers, connected and electric devices, sport articles such as tennis rackets, fishing rods, skis and many other "manufactured" products. All these industrial products could benefit from the embedded operational intelligence provided by the smart materials.

In the 21st Century, the development of products with increasing levels of functionalities by the use of smart materials will be more and more multisectoral. This will concern the following: transportation, agriculture, food and consumption packaging, construction, sports and leisure, white products and domestic products, healthcare, energy and environment, the Internet of Things, space and defense technologies and in the chemical industry sector opportunities of smart materials development and supply.

The concept of synthetic intelligent and biocompatible materials and devices could replace the inherent failing intelligence functions in our biological systems, as we grow older, for example. Emulating biological mechanisms will provide functionality and intelligence that will allow aging human beings to act in multiple environments. And finally, the development of synthetic smart systems in the form of robots not necessarily of humanoid type in the Living LAB (health care laboratory) could fully replace human functions in certain activities of the common domestic life or of the social and economic life in the industry, in the sector of strenuous work or health damaging tasks.

4.3. Thermodynamics of couplings in active materials

Piezoelectricity is an example of characteristics resulting from couplings between two different physical properties. Probably, the best-known phenomenon is the thermo-mechanical coupling associated with the thermal expansion of a material. It is used as an illustration for the introduction of the notion of couplings in a thermodynamic approach. This is followed by a description of other types of couplings underlying the functionalities described in relation with smart materials. Several applications of piezoelectricity are then presented, including the one already described in the context of technologies applied in near-field microscopy in Chapter 1.

4.3.1. *Thermo-mechanical and thermoelastic coupling*

When a beam is heated at one end, the heat or thermal energy gradually propagates and reaches the other end of the beam. This phenomenon is known as heat conduction. Heat losses should also be considered. These are due to the transfer from the surface of the material to the surroundings, in the form of convection, which is the transfer of thermal energy between a surface and a surrounding fluid (gas or liquid) or in the form of energy transfer by emission of radiation, which is in the infrared domain at ambient temperatures.

A material considered in a first approximation as a homogeneous body expands when its temperature increases. This phenomenon is modeled by a heat expansion coefficient α whose expression is given by:

$$\alpha = -\frac{1}{V}\left(\frac{\partial V}{\partial T}\right)_S \qquad [4.1]$$

where V is the volume of the body, T is the temperature, and S is the state function known as the entropy of the system. In this case, the thermodynamic state of the beam is assumed to move from a state of equilibrium characterized by an initial temperature T and a volume V, to an another state of equilibrium characterized by a temperature $T + dT$ and a volume $V + dV$, without any heat transfer to the external environment, which is known as an adiabatic process (isentropic if the process is reversible).

In the condensed phase, expansion is interpreted at the atomic level at the sub-nanometer scale by the average amplitude of vibration of atoms or ions in their crystallographic sites (anharmonic oscillator model, Chapter 3 of [DAH 16]) which increases when temperature rises. In order to take into account the symmetry properties of the crystallographic systems in which the material crystallizes at the considered temperature, it is more convenient to use tensor calculus [BRI 38, BRU 55, BIO 56, NYE 57, BRU 62, PRI 62, KIT 64, LAN 70] to model this thermo-mechanical effect. It is worth recalling that in a three-dimensional Euclidian space with three Cartesian axes (notation $1 = x$, $2 = y$ and $3 = z$) two-by-two orthogonal, the notion of tensor is related to the physical quantity represented by the dyadic product of two vectors. It is in the Riemann space where geodesics are curves, that the notion of tensor fully reveals its significance, particularly in the context of general relativity. Using the tensor notation, the temperature T being a scalar is a zero-order tensor (1 component), the force, which is a vector, is a first-order tensor (3 components) that is denoted by F_k and in order to represent the coupling between two vector quantities considered in the same system of axes, a second-order tensor (9 components) is used, which can be built by the dyadic product of two vectors and third-order tensors, fourth-order tensors, etc., can likewise be defined.

In order to describe the strain of a material, the hypothesis of a continuous medium is generally adopted. At any point of the material, represented by a vector r, small displacements represented by a translation vector $u(r)$ and the derivatives of the vector with respect to the coordinates are considered. When the material is deformed, at a point represented by r, of components x, y, z, each component of $u(r)$, $u_x(x, y, z)$, $u_y(x, y, z)$ and $u_z(x, y, z)$ is *a priori* subjected to different modifications along x, y and z. If the material is deformed, the variations of $u_x(x, y, z)$ along x, y and z, namely, the quantities $du_{xx}(x, y, z) = (\partial u_x / \partial x)dx$, $du_{xy}(x, y, z) = (\partial u_y / \partial y)dy$ and $du_{xz}(x, y, z) = (\partial u_z / \partial z)dz$ can be evaluated. Similarly, for the other two components of the vector $u(r)$, the various quantities lead to nine possible components for the mathematical representation of this strain.

The case of uniform strain along the three directions, when volume changes without a change in shape, is referred to as hydrostatic compression (or traction). If the shape changes while the volume remains unchanged, this is pure shearing. Nine derivatives are thus obtained, which are represented in the form of a strain tensor (second-order tensor) which can be split into a symmetric form:

$$\varepsilon_{ik} = \frac{1}{2}\left(\frac{\partial u_i}{\partial x_k} + \frac{\partial u_k}{\partial x_i}\right)$$

[4.2]

and an antisymmetric form:

$$\omega_{ik} = \frac{1}{2}\left(\frac{\partial u_i}{\partial x_k} - \frac{\partial u_k}{\partial x_i}\right)$$

[4.3]

which represent, respectively, the deformation of the material at the point and the rotation of the material about this point.

Figure 4.6 shows a schematic representation of various types of displacements in a material in the context of the study of strains ($\boldsymbol{u}(B_0)$) in a continuous medium that can be split into global translational motion ($\boldsymbol{u}(A_0)$), rotational motion ($\boldsymbol{\omega}(A_0)\wedge d\boldsymbol{r}$) and symmetric strain ($\boldsymbol{\varepsilon}(A_0)d\boldsymbol{r}$) (or vibrations if dynamics is included) where $((u(B_0)) = (u(A_0)) + (\boldsymbol{\omega}(A_0)\wedge d\boldsymbol{r}) + (\boldsymbol{\varepsilon}(A_0)d\boldsymbol{r})$.

In a one-dimensional medium, such as a wire, strain mainly occurs along the axis of the wire. In this case $i=1$ and $k=1$, and there is only one term that is ε_{11}, which is denoted by ε. In the case of a plane surface, there are only two possibilities for i and k, so that four terms are to be considered.

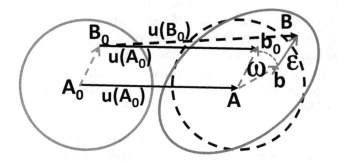

Figure 4.6. *The various displacements of the translational vector u(r) in a material in terms of global translation, rotation and symmetric deformation. For a color version of this figure, see www.iste.co.uk/dahoo/metrology1.zip*

Consider as an exercise the calculation of the strain tensor for a 2D material whose thickness is negligible compared to the other two dimensions, as in the thin film typology discussed in Chapter 1.

The expression of the thermal expansion for a homogeneous variation ΔT of the temperature of the material is:

$$\varepsilon_{ik} = \alpha_{ik}\,\Delta T \qquad [4.4]$$

where α_{ik} are the coefficients of the symmetric tensor of thermal expansion $[\alpha_{ik}]$ and ε_{ik} are the coefficients of the strain tensor $[\varepsilon_{ik}]$.

This relation can be written in a matrix form as:

$$\begin{pmatrix} \varepsilon_{11} & \varepsilon_{12} & \varepsilon_{13} \\ \varepsilon_{12} & \varepsilon_{22} & \varepsilon_{23} \\ \varepsilon_{13} & \varepsilon_{23} & \varepsilon_{33} \end{pmatrix} = \begin{pmatrix} \alpha_{11} & \alpha_{12} & \alpha_{13} \\ \alpha_{12} & \alpha_{22} & \alpha_{23} \\ \alpha_{13} & \alpha_{23} & \alpha_{33} \end{pmatrix}(T - T_0) \qquad [4.5]$$

Given that the strain tensor is symmetric, only six of the nine terms of the tensor are independent. In this case, Voigt (or Voigt-Nye) notation can be used to transform the matrix representation to a vector representation as follows:

$$\begin{pmatrix} \varepsilon_{11} & \varepsilon_{12} & \varepsilon_{13} \\ \varepsilon_{12} & \varepsilon_{22} & \varepsilon_{23} \\ \varepsilon_{13} & \varepsilon_{23} & \varepsilon_{33} \end{pmatrix} \equiv \begin{pmatrix} \varepsilon_{11} \\ \varepsilon_{22} \\ \varepsilon_{33} \\ \varepsilon_{23} \\ \varepsilon_{13} \\ \varepsilon_{12} \end{pmatrix} = \begin{pmatrix} \varepsilon_1 \\ \varepsilon_2 \\ \varepsilon_3 \\ \dfrac{\varepsilon_4}{2} \\ \dfrac{\varepsilon_5}{2} \\ \dfrac{\varepsilon_6}{2} \end{pmatrix} = \begin{pmatrix} \alpha_1 \\ \alpha_2 \\ \alpha_3 \\ \dfrac{\alpha_4}{2} \\ \dfrac{\alpha_5}{2} \\ \dfrac{\alpha_6}{2} \end{pmatrix}(T - T_0) \qquad [4.6]$$

where (ε) and (α) are the vectors with six lines (6x1), and $(T - T_0)$ is a scalar.

On the proper axes of the expansion tensor, the following relation can be written as:

$$\varepsilon_i = \alpha_i\,\Delta T \qquad [4.7]$$

where α_i is the main expansion coefficients, $i=1,2$ and 3. In this case, the matrix form is written as:

$$\begin{pmatrix} \varepsilon_1 & 0 & 0 \\ 0 & \varepsilon_2 & 0 \\ 0 & 0 & \varepsilon_3 \end{pmatrix} = \begin{pmatrix} \alpha_1 & 0 & 0 \\ 0 & \alpha_2 & 0 \\ 0 & 0 & \alpha_2 \end{pmatrix} (T - T_0) \tag{4.8}$$

and the vector form as:

$$\begin{pmatrix} \varepsilon_1 \\ \varepsilon_2 \\ \varepsilon_3 \\ 0 \\ 0 \\ 0 \end{pmatrix} = \begin{pmatrix} \alpha_1 \\ \alpha_2 \\ \alpha_3 \\ 0 \\ 0 \\ 0 \end{pmatrix} (T - T_0) \tag{4.9}$$

To model the effects of a mechanical stress in a material according to the formalism of the mechanics of continuous media or the theory of elasticity, elementary volumes are considered for the establishment of the constituent relations. The latter are infinitely small at the macroscopic scale, but very large at the scale of the constituents of the material (atoms, molecules, etc.). During a mechanical deformation of the medium, in the infinitesimal volume around each point, the forces responsible for internal stress are assumed to be short-range and limited to near neighbors. In this case, these forces are exerted by one of its elements on the neighboring elements so that in a given volume inside the material, they can only act at the surface.

The establishment of the constitutive equations requires the definition of the stress tensor acting on an elementary volume of the material. The stress tensor σ_{ik} describes the state of stress at any point of the volume considered in the material and in all the directions. Its components that correspond to a force exerted on a unit surface around a point are homogeneous to a pressure (in Pa or $N.m^{-2}$). It is a symmetric tensor such that $\sigma_{ik} = \sigma_{ki}$. The diagonal terms of the stress tensor are given by tensile (or compression) stresses and non-diagonal terms by shearing stresses. The components have positive values in the case of tension (negative in the case of compression).

In Figure 4.7, the components of the stress tensor are represented by its indices on each face identified by the normal vector n_i.

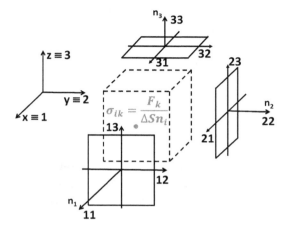

Figure 4.7. *Stress tensor σ_{ik} corresponding to the component k of the force F per unit area acting on the point in the middle of an area defined by its normal n_i. For a color version of this figure, see www.iste.co.uk/dahoo/metrology1.zip*

The matrix form of tensor σ_{ik} is given by:

$$\begin{pmatrix} \sigma_{xx} & \sigma_{xy} & \sigma_{xz} \\ \sigma_{yx} & \sigma_{yy} & \sigma_{yz} \\ \sigma_{zx} & \sigma_{zy} & \sigma_{zz} \end{pmatrix} = \begin{pmatrix} \sigma_{11} & \sigma_{12} & \sigma_{13} \\ \sigma_{12} & \sigma_{22} & \sigma_{23} \\ \sigma_{13} & \sigma_{23} & \sigma_{33} \end{pmatrix} \qquad [4.10]$$

Given that the stress tensor is symmetric, similarly to the strain tensor, only six of nine terms of the tensor are independent. Voigt (or Voigt-Nye) notation can be used to transform the matrix representation to the following vector notation:

$$\begin{pmatrix} \sigma_{11} & \sigma_{12} & \sigma_{13} \\ \sigma_{12} & \sigma_{22} & \sigma_{23} \\ \sigma_{13} & \sigma_{23} & \sigma_{33} \end{pmatrix} \equiv \begin{pmatrix} \sigma_{11} \\ \sigma_{22} \\ \sigma_{33} \\ \sigma_{23} \\ \sigma_{13} \\ \sigma_{12} \end{pmatrix} = \begin{pmatrix} \sigma_1 \\ \sigma_2 \\ \sigma_3 \\ \sigma_4 \\ \sigma_5 \\ \sigma_6 \end{pmatrix} \qquad [4.11]$$

where (σ) is the vector with six lines (6x1). The first three terms (σ_1, σ_2, σ_3) are the tensile or compression stresses along the perpendicular to surfaces, hence parallel to axes 1, 2 and 3 defined by the normal vector to the surface and the last three terms ($\sigma_4, \sigma_5, \sigma_6$) are the shear stresses about axes 1, 2 and 3 or torques in planes 2-3, 1-3 and 1-2 (Figure 4.7).

The resultant of these forces is given by a volume integral that can be expressed as a surface integral as follows:

$$\iiint_{volume} F_i dV = \iiint_{volume} \frac{\partial \sigma_{ik}}{\partial x_k} dV = \oiint \sigma_{ik} dA_k \qquad [4.12]$$

where F_i is the ith component of the force acting on the volume, $\partial \sigma_{ik}/\partial x_k$ is the divergence of the stress tensor $[\sigma_{ik}]$, dA_k is the elementary surface oriented along the normal defined by the unit vector n_k and where the integral is evaluated on the closed surface surrounding the volume V. Since, by definition, $\sigma_{ik} = \sigma_{ki}$ is the ith component of the force acting on a unit area perpendicular to the axis of x_k (Figure 4.7). Equation [4.10] shows that $\sigma_{ik} dA_k$ is the ith component of the force acting on the area dA.

When a material is subjected to a uniform compression (pressure forces P), known as hydrostatic, then:

$$\sigma_{ik} = -P \delta_{ik} \qquad [4.13]$$

where δ_{ik} is the Kronecker symbol (0 if $i \neq k$, 1 otherwise).

At equilibrium, a strained material verifies (according to the Einstein convention involving summation when indices are repeated) the following relation:

$$\frac{\partial \sigma_{ik}}{\partial x_k} = 0 \qquad [4.14]$$

For small strains, the material returns to its initial unstrained state when external forces are released. In this case, it undergoes elastic strains, and the stored energy density is a state function. On the contrary, if the strain is permanent, the transformation is termed as plastic. It is worth noting that in certain situations, this strain can be viscoplastic.

In the presence of an elastic strain, the variation of internal energy according to the first law of thermodynamics is written as:

$$dU = TdS - \sigma_{ik}d\varepsilon_{ik} \qquad [4.15]$$

where the first term corresponds to an exchange in the form of heat, and the second term corresponds to an exchange in the form of mechanical work.

If the applied force is a uniform compression (equation 4.13), the result is the standard expression of the first law of thermodynamics, which reflects the conservation of energy of the system represented by the material such that:

$$dU = TdS - PdV \qquad [4.16]$$

The result is, $\delta_{ik}d\varepsilon_{ik}$ being equal to $d\varepsilon_{ii}$, the trace of tensor $[d\varepsilon_{ik}]$. The latter is an invariant that corresponds to the elementary volume dV.

In equation 4.16, T is the intensive variable temperature (an average value over the whole material in equilibrium) and S is the entropy state function, which is an extensive variable (that doubles when volume doubles). Only two of the four laws of thermodynamics are applied, namely, the first law of energy conservation and the second law on the irreversibility of physical processes that stipulates that the entropy of an isolated system always increases. The variation of entropy is defined from the amount of energy exchanged in the form of heat. This variation is zero for a reversible transformation and positive for an irreversible transformation, which is expressed by:

$$dS = \left(\frac{\delta Q}{T}\right)_{rev} + \left(\frac{\delta Q}{T}\right)_{irrev} \qquad [4.17]$$

where δQ is the energy supplied as heat. This form of energy modifies neither the external parameters of the system that receives or releases the heat nor the configuration of the system external environment. On the contrary, the energy exchanged in the form of work involves either a modification of the external parameters of the system that receives or releases the work or the configuration of the system external environment or both.

The coupling between two types of properties is the channel through which the energy supplied in one form is transformed in another form. It is appropriate to study the physical phenomena resulting from possible couplings through thermodynamics. The procedure implies identifying the appropriate state function to evaluate this energy transfer from one physical characteristic to another.

Based on equation [4.15], entropy variation can be expressed as:

$$dS = \frac{1}{T}dU + \frac{\sigma_{ik}}{T}d\varepsilon_{ik} \qquad [4.18]$$

A comparison between equations [4.17] and [4.18] reveals the equivalence between the purely thermal form of energy and the mechanical form. Expression [4.15] or [4.18] being convenient for adiabatic processes ($S = 0$), Helmholtz free energy, $F = U\text{-}TS$ that is obtained from U by the Legendre transform is preferable in the case of isothermal processes ($T = 0$) as it can be written as:

$$dF = -SdT - \sigma_{ik}d\varepsilon_{ik} \qquad [4.19]$$

Equations [4.15] and [4.19] show that the stress tensor is expressed in the form of a derivative of U at constant entropy or the derivative of F at constant temperature such that:

$$\sigma_{ik} = \left(\frac{\partial U}{\partial \varepsilon_{ik}}\right)_S = \left(\frac{\partial F}{\partial \varepsilon_{ik}}\right)_T \qquad [4.20]$$

The strain tensor can also be expressed as the partial derivative of another potential function. Indeed, choosing by means of another Legendre transform, the thermodynamic potential as $\Phi = F\text{-}\sigma_{ik}\varepsilon_{ik}$, it leads to:

$$d\Phi = -SdT - \varepsilon_{ik}d\sigma_{ik} \qquad [4.21]$$

So that ε_{ik} is determined in the following form:

$$\varepsilon_{ik} = -\left(\frac{\partial \Phi}{\partial \sigma_{ik}}\right)_T \qquad [4.22]$$

Starting from the power series expansion of the Helmholtz free energy, F, or of the internal energy, U, in terms of ε_{ik}, the stress tensor can be expressed as a function of the strain tensor for an isothermal and isentropic strain, respectively, in the following form:

$$\sigma_{ik} = \left(\frac{\partial^2 F}{\partial \varepsilon_{ik} \partial \varepsilon_{lm}} \right)_T \varepsilon_{lm} \text{ or } \sigma_{ik} = \left(\frac{\partial^2 U}{\partial \varepsilon_{ik} \partial \varepsilon_{lm}} \right)_S \varepsilon_{lm} \qquad [4.23]$$

These relations are generally expressed in the following form:

$$\sigma_{ik} = C^T_{iklm} \varepsilon_{lm} \text{ or } \sigma_{ik} = C^S_{iklm} \varepsilon_{lm} \qquad [4.24]$$

where C_{iklm} are the fourth-order tensors that correspond to isothermal or isentropic stiffness constants. This form corresponds to Hooke's law, which governs linear elasticity.

Likewise, Hooke's law can be expressed in the inverse form:

$$\varepsilon_{ik} = S^T_{iklm} \sigma_{lm} \text{ or } \varepsilon_{ik} = S^S_{iklm} \sigma_{lm} \qquad [4.25]$$

where S_{iklm} are compliance constants.

Given that each index can take three values, there are *a priori* $3^4 = 81$ constants. The constants must verify the following symmetry relations: $C_{iklm} = C_{kilm} = C_{ikml} = C_{lmik}$. Indeed, the strain and stress tensors being symmetric ($C_{iklm} = C_{kilm} = C_{ikml}$), there are only 36 possibilities left (the (3+2+1) pairs of indices raised to the power of 2 or 6^2). Furthermore, as the second derivatives verify the Cauchy–Schwarz relation ($C_{iklm} = C_{lmik}$), there are only 21 possibilities (6+5+4+3+2+1) left. It is worth noting that depending on the symmetry elements of a crystal (Appendix and Tables 4.8 and 4.9), this number can be smaller.

The indices of the fourth-order tensor can be contracted, applying the same procedure as for the second-order tensors, ε_{ik} and σ_{ik} using the rule shown in Tables 4.2 and 4.3 for the correspondences between the partitioning (A, B, etc.) of the indices *iklm* of the fourth-order tensor C_{iklm} and those of the indices ik of the second-order tensor C_{ik}.

C_{iklm}	ik	11 22 33	23 31 12	32 13 21
lm				
11 22 33		A	B	C
23 31 12		E	F	G
32 13 21		H	I	J

Table 4.2. *Correspondence for contracting indices C_{iklm} to C_{ik}*

C_{ik}	i	1 2 3	4 5 6
k			
1 2 3		A	B & C
4 5 6		E & H	F & G & I & J

Tables 4.3. *Correspondence for contracting indices C_{iklm} to C_{ik}*

Relations [4.24] and [4.25] can thus be written as follows:

$$\sigma_i = C_{ik}\varepsilon_k \qquad\qquad [4.26]$$

$$\varepsilon_i = S_{ik}\sigma_k \qquad\qquad [4.27]$$

where C_{ik} and S_{ik} are 6 x 6 matrices.

The strain and stress tensors being second-order symmetric tensors, they can be decomposed into a spherical part and a deviatory part as follows:

$$\varepsilon_{ik} = (\varepsilon_{ik} - \frac{1}{3}\delta_{ik}\varepsilon_{ll}) + \frac{1}{3}\varepsilon_{ll} \qquad [4.28]$$

$$\sigma_{ik} = (\sigma_{ik} - \frac{1}{3}\delta_{ik}\sigma_{ll}) + \frac{1}{3}\sigma_{ll} \qquad [4.29]$$

The first term of equation [4.28] corresponds to a sliding or shearing term (strain without change of volume) and the second term to uniform compression (strain without change of shape). The free energy F can then be expressed as follows:

$$F - F_0 = \mu(\varepsilon_{ik} - \frac{1}{3}\delta_{ik}\varepsilon_{ll})^2 + (\lambda + \frac{2}{3}\mu)\varepsilon_{ll}^2 \qquad [4.30]$$

$$F - F_0 = \mu(\varepsilon_{ik} - \frac{1}{3}\delta_{ik}\varepsilon_{ll})^2 + \frac{K}{2}\varepsilon_{ll}^2 \qquad [4.31]$$

where F_0 is the free energy in the absence of strain, and λ and μ are Lamé coefficients. K is the bulk modulus or rigidity modulus, and μ is the shear modulus ($K > 0$ and $\mu > 0$). These moduli can also be expressed as follows:

$$\mu = \frac{E}{2(1+v)} \text{ and } K = \frac{E}{3(1-2v)} \qquad [4.32]$$

where E is Young's modulus and v is Poisson's ratio.

The expression of free energy for strains resulting from a variation of temperature is given by:

$$F(T) - F(T_0) = -K\alpha(T - T_0)\varepsilon_{ll} + \mu(\varepsilon_{ik} - \frac{1}{3}\delta_{ik}\varepsilon_{ll})^2 + \frac{K}{2}\varepsilon_{ll}^2 \qquad [4.33]$$

where $F(T_0)$ is the free energy in the absence of strain at temperature T_0, and α is the thermal expansion coefficient of the material considered isotropic.

If the nine components of the stress tensor σ_{ik} and the temperature T are considered as independent variables, the variations of the strain tensor components ε_{ik} and of the entropy state function S can be expressed as a function of these independent variables, σ_{ik} and T. Similarly, the components of the strain tensor ε_{ik} and the entropy state function S can be taken as independent variables and the variations of the components of the stress tensor σ_{ik} and of temperature T can be expressed as a function of ε_{ik} and S.

The following equations are obtained with the first set of variables:

$$d\varepsilon_{ik} = \left(\frac{\partial \varepsilon_{ik}}{\partial \sigma_{lm}}\right)_T d\sigma_{lm} + \left(\frac{\partial \varepsilon_{ik}}{\partial T}\right)_\sigma dT \qquad [4.34]$$

$$dS = \left(\frac{\partial S}{\partial \sigma_{lm}}\right)_T d\sigma_{lm} + \left(\frac{\partial S}{\partial T}\right)_\sigma dT \qquad [4.35]$$

The physical significance of these four partial derivatives can be easily established with the following approach, by considering the variations of ε_{ik} and S, when $dT=0$ (isothermal) and $d\sigma_{lm}=0$ (constant stress). The first term of [4.34] corresponds to the isothermal elastic coefficients as already obtained with the free energy function F and denoted: $\left(\frac{\partial \varepsilon_{ik}}{\partial \sigma_{lm}}\right)_T = S_{iklm}^T$ (equation [4.25]). The superscript designates the parameter that is constant. The second term represents the thermal expansion coefficients: $\alpha_{ik} = \left(\frac{\partial \varepsilon_{ik}}{\partial T}\right)_\sigma$ (equation [4.44]). The first term of equation [4.35] corresponds to the entropy increase following stresses in an isothermal process. This term corresponds to the ratio of the heat generated as result of these stresses and the temperature T, hence $\left(\frac{\partial S}{\partial \sigma_{lm}}\right)_T = \frac{\delta Q}{T}$, which is the piezo-caloric effect. If the second term of equation [4.35] is multiplied by T, hence $T\left(\frac{\partial S}{\partial T}\right)_\sigma = C^\sigma$, the result corresponds to the heat released when the stress is maintained constant (equivalent to a constant pressure) which corresponds to the heat capacity C^σ per unit volume under constant stress. Given the interactions between the effects related to these coefficients, the latter are not independent.

Based on the potential Φ, and on the variation $d\Phi$ (equation [4.21]), the following relation can be written as:

$$\frac{\partial^2 \Phi}{\partial \sigma_{lm} \partial T} = -\left(\frac{\partial S}{\partial \sigma_{lm}}\right)_T = -\left(\frac{\partial \varepsilon_{lm}}{\partial T}\right)_\sigma \qquad [4.36]$$

which shows that there is a relation between the thermal expansion coefficient and the piezo-caloric coefficient. For small variations, the integration of equations [4.34] and [4.35] leads to the following equations:

$$\varepsilon_{ik} = S^T_{iklm}\sigma_{lm} + \alpha_{ik}\Delta T \qquad [4.37]$$

$$\Delta S = \alpha_{ik}\sigma_{ik} + \frac{c^\sigma}{T}\Delta T \qquad [4.38]$$

After contraction of the indices, these equations between tensors are simplified and can be written as:

$$\varepsilon_i = S^T_{ik}\sigma_k + \alpha_i\Delta T \qquad [4.39]$$

$$\Delta S = \alpha_i\sigma_i + \frac{c^\sigma}{T}\Delta T \qquad [4.40]$$

These equations can be written in the form of matrix equations as:

$$\varepsilon = S^T\sigma + \alpha\Delta T \qquad [4.41]$$

$$\Delta S = \alpha\sigma + \frac{c^\sigma}{T}\Delta T \qquad [4.42]$$

It is worth noting that in the presence of a heat flux, the heated region cannot freely expand and that it is subjected to compression stresses of the colder regions. Similarly, the colder regions are subjected to tensile stresses exerted by the warmer regions. In this case, these additional stresses must be taken into account when applying Hooke's law.

It can thus be shown that

$$\sigma_{ik} = -3K\alpha(T - T_0) + 2\mu\varepsilon_{ik} + \frac{E\nu}{(1+\nu)(1-2\nu)}\varepsilon_{ll}\delta_{ik} \qquad [4.43]$$

and

$$\varepsilon_{ik} = \alpha(T - T_0)\delta_{ik} + \frac{1+\nu}{E}\sigma_{ik} - \frac{\nu}{E}\sigma_{ll}\delta_{ik} \qquad [4.44]$$

These equations are applied to determine the strain when temperature varies. In the weak coupling mode, the thermal effect is first evaluated, and from the new set of values of the different parameters, the mechanical effect is then evaluated and so on in a feedback loop until the final thermodynamic state is reached. In the strong coupling mode, both effects must be considered simultaneously.

4.3.2. *Multiphysics couplings*

The connection between the electric and magnetic phenomena was established in the 19th Century, when Oersted discovered that an electric current generates a magnetic field. This effect was observed by the deviation of a compass needle when an electric current starts or stops flowing through a conductor during a Joule effect didactic experiment. A thermoelectric effect leads to a temperature increase, when an electric current flows through a metallic material. The electromagnetic properties of matter are described in terms of electric polarization and magnetization, locally defined as densities of dipolar electric moments and dipolar magnetic moments in the context of Maxwell's equations by assuming that the material that is polarized when subjected to an electric or magnetic field is in the form of a continuous medium. Table 4.4 summarizes the fundamental relations of electromagnetism.

Medium	Isotropic	Anisotropic	Constant
Dielectric	$D=\varepsilon E$ $P=\varepsilon_0\chi_e E$ $div\varepsilon_0 E=-divP$	$D=\varepsilon E$ $P=\varepsilon_0\chi_e E$ $div\varepsilon_0 E=-divP$	Permittivity Electric susceptibility
Magnetic	$B=\mu H$ $J=\mu_0\chi_m H$ $rot\mu_0 H=rotJ$	$B=\mu H$ $J=\mu_0\chi_m H$ $rot\mu_0 H=rotJ$	Permeability Magnetic susceptibility
Conductor	$E=\rho_e j$ $j=\sigma_e E$	$E=\rho_e j$ $J=\sigma_e E$	Resistivity Electrical conductivity
Superconductor	$dj/dt=(1/\mu_0\lambda^2_s)E$ $j=-(1/\mu_0\lambda^2_s)A$		Skin depth

Table 4.4. *Fundamental laws for linear electromagnetic media*

Most of the discoveries concerning functional materials characterized by a coupling between two types of properties can also be traced back to the 19th Century. The origin and the comprehension of the coupling between various physical properties within the same material is still an active research subject in condensed matter physics. The existence of materials with couplings that involve several properties simultaneously has been demonstrated. In the ferroics, for example, the coupling depends on the shaping of the material, on the crystallographic structure and on the configuration of the magnetic spins. These materials are characterized by properties such as ferromagnetism, ferroelectricity and/or ferroelasticity.

A thermoelectric effect of another kind was revealed by Thomas Johann Seebeck in 1821. He discovered that a metallic needle is deviated when it is placed between two conductors of different kinds connected by junctions at their ends and subjected to a thermal gradient (Seebeck effect). The observed effect has an electric origin, although Seebeck first associated the cause with a magnetic field. An electric potential difference appears at the junction between the two different materials subjected to a temperature difference. The most widely known application of the Seebeck effect is the measurement of temperature using thermocouples. The pyroelectric effect studied by Antoine Becquerel in 1828 is another thermo-electric effect. The direct effect is the appearance of polarization or charges under thermal excitation, while the inverse effect is the electro-caloric effect, with the emission of heat or a warming effect under electric excitation. An example of such a material is lithium tetra-borate ($Li_2B_4O_7$) in which a change in temperature leads to a variation of the electric polarization of the crystal. Finally, in 1834, Jean-Charles Peltier discovered another thermoelectric effect: a temperature difference appears at the junction between two different materials subjected to an electric current (see the Peltier effect). Between 1878 and 1893, Lord Kelvin developed the theory leading to the interpretation of the properties of pyroelectric materials and of the Seebeck and the Peltier effects. In a material subjected to a temperature gradient and through which an electric current flows, heat exchanges occur with the external environment. Conversely, an electric current is generated by a material through which there is a heat flux. The Thomson effect, the Seebeck effect and the Peltier effect show similarities. However, the Thomson effect is different from the other ones because only one conductive material is necessary for its occurrence.

Magnetostriction was discovered in 1842 by the English physicist Joule when an iron beam was placed in a magnetic field. Electrostriction was predicted and verified by Jacques and Pierre Curie in 1881 on a quartz crystal. It is an electromechanical effect that depends on the squared of the intensity of the electric field inducing the material strain. This is known as an electrostriction effect and is generally negligible.

The piezoelectric effect is another electromechanical effect that was experimentally discovered in the tourmaline crystal (hemihedral crystals with inclined faces) by the Curie brothers, Jacques and Pierre, in 1880 [CUR 80]. Combining their knowledge on pyroelectricity and on the crystalline structure, they predicted and verified the existence of piezoelectricity in materials such as quartz, tourmaline, topaz, sugar and Rochelle salt crystals, which become polarized under the effect of a pressure. The possibility of the reverse effect was predicted the following year by Gabriel Lippmann [LIP 81, LIP 81b] as a result of thermodynamic calculations and was verified the same year by the Curie brothers [CUR 81]. In the same year 1881, Wilhelm Hankel used the term piezoelectricity originating in the Greek word "piezein", meaning to press or push, to qualify this phenomenon. In 1910, the theory of this effect was described in the context of tensor formalism and crystallography by Woldemar Voigt [VOI 10, VOI 28]. He listed the 20 crystalline classes (absence of a center of symmetry) that can present piezoelectricity and specified the piezoelectric constants according to the crystalline structures. In a first approximation piezoelectricity is a linear effect as the strain is proportional to the applied electric field. The electromechanical coupling coefficient, which is the essential parameter for the description of piezoelectric materials, reflects the conversion of electric energy into mechanical energy and vice versa.

Piezoelectric materials are available in various forms:

– Monocrystals: this is the form of the piezoelectric materials found in nature, such as the insulating dielectrics quartz or tourmaline, the semiconductors CdS or AsGa, the oxide ZnO or the ferroelectric monocrystal LiNbO.

– Ferroelectric ceramics ($BaTiO_3$ or $Pb(Zr, Ti)O_3$), based on a material with perovskite structure (Figure 4.8), which are the most widely used, mainly due to their ease of fabrication and the many properties conferred by the perovskite structure when the chemical composition and the manufacturing parameters are varied.

– Composites, whose electromechanical coupling coefficient is higher than that of conventional ceramics.

– Polymers (polyvinylidene fluoride PVDF, polyvinyl chloride PVC)), whose ease of fabrication allows a variety of elaboration process and have the advantage of being very flexible. Their coupling coefficients are nevertheless quite low.

– Biological materials, such as cellulose, amylose, keratin, polypeptides and polyaminoacids.

Perovskite is a natural mineral with the chemical formula $CaTiO_3$. It is the prototype of many materials whose structure is of the type ABO_3 (Figure 4.8). This structure is very stable and enables the manufacturing of a very large number of components with a wide variety of properties that can be used in many practical applications. The key role of the BO_6 octahedrons is to confer to the crystal the property of being ferromagnetic or/and ferroelectric. These materials are formed of oxygen octahedrons linked by their vertices their centers being the sites where cations B are located. Cations A, of larger size, are located in the cuboctahedric cavity formed by eight octahedrons. In this ideal structure, a factor t or a Goldsmith factor can be defined in relation to the radii of various ions. When it is equal to 1, the prototype structure is stable.

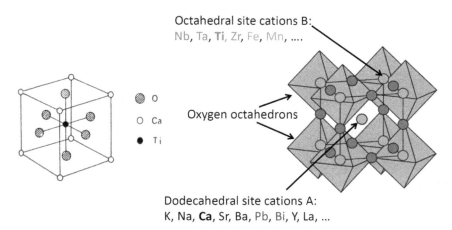

Figure 4.8. *Octahedral and dodecahedral sites of $CaTiO_3$ perovskite with ABO_3 structure. For a color version of this figure, see www.iste.co.uk/dahoo/metrology1.zip*

The crystals whose unit cell has no center of symmetry (non-centrosymmetric) have all piezoelectric properties with the exception of crystals described by 432 point symmetry. Piezoelectric crystals can be classified into 20 point groups (32 groups – 11 centrosymmetric – 1). However, the appearance of piezoelectricity requires the application of an electric or mechanical external stress in a given direction, as the effect is not isotropic.

The direct piezoelectric effect can be described by means of a third rank tensor denoted by d_{ikl}, which links polarization to mechanical stress σ_{kl} such as:

$$D_i = d_{ikl}\sigma_{kl} \qquad [4.45]$$

The stress tensor being symmetric, the number of independent coefficients is reduced from 27 (3^3) to 18 (3x3x2 or 3x6, or 3 possibilities for i and 6 for kl). If the crystal has symmetry elements, this number is lowered and reduces to 4 in quartz and to 3 in barium titanate ($BaTiO_3$).

Indices of third-order tensor can also be contracted by considering the symmetry properties according to the Voigt convention shown in Table 4.5.

d_{ikl}	i11	i22	i33	i23 i32	i31 i13	i12 i21
d_{ik}	i1	i2	i3	i4	i5	i6

Table 4.5. *Correspondence for contracting the indices from d_{ikl} to d_{ik}*

A relation in the following matrix form is obtained as:

$$D_i = d_{ik}\sigma_k \qquad [4.46]$$

The converse piezoelectric effect can be described using a third rank tensor that is also denoted by d_{ikl} which links the strain to the electric field E_k:

$$\varepsilon_{ik} = d_{ikl}E_k \qquad [4.47]$$

The contraction of indices leads to a matrix relation that can be written in the following form:

$$\varepsilon_i = d_{ik}E_k \qquad [4.48]$$

It can be shown [NYE 57] that the piezoelectric coefficients of the direct and converse effect are equal provided that the following correspondences are used during the contraction of indices:

$$\begin{pmatrix} \sigma_{11} & \sigma_{12} & \sigma_{13} \\ \sigma_{21} & \sigma_{22} & \sigma_{23} \\ \sigma_{31} & \sigma_{32} & \sigma_{33} \end{pmatrix} = \begin{pmatrix} \sigma_1 & \sigma_6 & \sigma_5 \\ \sigma_6 & \sigma_2 & \sigma_4 \\ \sigma_5 & \sigma_4 & \sigma_3 \end{pmatrix} \equiv \begin{pmatrix} \sigma_1 \\ \sigma_2 \\ \sigma_3 \\ \sigma_4 \\ \sigma_5 \\ \sigma_6 \end{pmatrix} \qquad [4.49]$$

$$\begin{pmatrix} \varepsilon_{11} & \varepsilon_{12} & \varepsilon_{13} \\ \varepsilon_{21} & \varepsilon_{22} & \varepsilon_{23} \\ \varepsilon_{31} & \varepsilon_{32} & \varepsilon_{33} \end{pmatrix} = \begin{pmatrix} \varepsilon_1 & \frac{1}{2}\varepsilon_6 & \frac{1}{2}\varepsilon_5 \\ \frac{1}{2}\varepsilon_6 & \varepsilon_2 & \frac{1}{2}\varepsilon_4 \\ \frac{1}{2}\varepsilon_5 & \frac{1}{2}\varepsilon_4 & \varepsilon_3 \end{pmatrix} \equiv \begin{pmatrix} \varepsilon_1 \\ \varepsilon_2 \\ \varepsilon_3 \\ \frac{1}{2}\varepsilon_4 \\ \frac{1}{2}\varepsilon_5 \\ \frac{1}{2}\varepsilon_6 \end{pmatrix} \qquad [4.50]$$

with

$$\begin{pmatrix} d_{11} & d_{12} & d_{13} & d_{14} & d_{15} & d_{16} \\ d_{21} & d_{22} & d_{23} & d_{24} & d_{25} & d_{26} \\ d_{31} & d_{32} & d_{33} & d_{34} & d_{35} & d_{36} \end{pmatrix} =$$
$$\begin{pmatrix} d_{111} & d_{122} & d_{133} & \frac{d_{123}}{2} & \frac{d_{131}}{2} & \frac{d_{112}}{2} \\ d_{211} & d_{222} & d_{233} & \frac{d_{223}}{2} & \frac{d_{231}}{2} & \frac{d_{212}}{2} \\ d_{311} & d_{322} & d_{333} & \frac{d_{323}}{2} & \frac{d_{331}}{2} & \frac{d_{312}}{2} \end{pmatrix} \qquad [4.51]$$

Table 4.6 summarizes the equations defining the direct and converse piezoelectric effects.

	Tensor notation	Matrix notation
Direct effect	$D_i = d_{ikl}\sigma_{kl}$	$D_i = d_{ik}\sigma_k$
Converse effect	$\varepsilon_{ik} = d_{lik}E_l$	$\varepsilon_i = d_{li}E_l$

Table 4.6. *Piezoelectric effect in tensor and matrix notation*

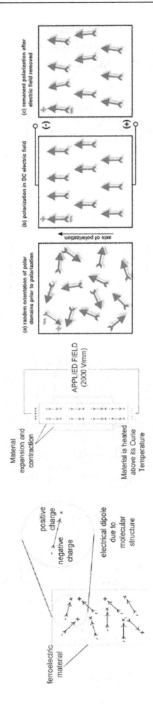

Figure 4.9. *Poling of a ferroelectric material. For a color version of this figure, see www.iste.co.uk/dahoo/metrology1.zip*

The family of piezoelectric ceramics forms an important group of materials. There are polycrystalline ferroelectric materials, most of which have a tetragonal/rhombohedral structure, close to the cubic structure. There are also mixtures of lead oxide, the most common of which is PZT. Unlike quartz, for example, these materials must be polarized in order to exhibit a piezoelectric effect. On the contrary, the latter is much stronger. Typically, a PZT beam subjected to an electric field presents a strain that is several hundred times greater than that of a quartz beam of similar size subjected to the same field.

The process used for the fabrication of piezoelectric materials is known as poling. The procedure for poling a ferroelectric material involves the alignment of electric dipoles that are randomly oriented in the material in the direction of an electric field when it is maintained at a temperature above its transition temperature. During the alignment, the material undergoes an elongation along the direction of the electric field and a contraction of its lateral dimensions. Then while maintaining the electric field the temperature is lowered (field cool). The material thus maintains a residual macroscopic polarization, the dipoles being nearly aligned even below the transition temperature after the removal of the electric field. The process is schematically represented in Figure 4.9.

As mentioned in the previous section for the thermo-mechanical effect and the piezo-caloric coupling, a system at equilibrium can be fully characterized if the extensive and intensive variables that determine the state functions are known, relying on the laws of thermodynamics. For piezoelectric materials, the quantities used are the entropy, the strain and the polarization of the material system, which are variables of a thermodynamic potential. Gibbs free energy is generally used, which is a function of intensive quantities: temperature, electric field and stress. It is expressed as:

$$\Phi = U - \sigma_{kl}\varepsilon_{kl} - E_k D_k - TS \qquad [4.52]$$

hence:

$$d\Phi = -\varepsilon_{kl}d\sigma_{kl} - D_k dE_k - SdT \qquad [4.53]$$

with

$$-\left(\frac{\partial^2 \Phi}{\partial \sigma_{kl} \partial E_i}\right)_T = \left(\frac{\partial \varepsilon_{kl}}{\partial E_i}\right)_{\sigma,T} = -\left(\frac{\partial D_i}{\partial \sigma_{kl}}\right)_{E,T} = d_{ikl}^T \qquad [4.54]$$

$$-\left(\frac{\partial^2 \Phi}{\partial \sigma_{kl} \partial T}\right)_E = \left(\frac{\partial \varepsilon_{kl}}{\partial T}\right)_{\sigma,E} = -\left(\frac{\partial S}{\partial \sigma_{kl}}\right)_{E,T} = \alpha_{kl}^E \qquad [4.55]$$

$$-\left(\frac{\partial^2 \Phi}{\partial E_i \partial T}\right)_\sigma = \left(\frac{\partial S}{\partial E_i}\right)_{\sigma,T} = -\left(\frac{\partial D_i}{\partial T}\right)_{\sigma,E} = p_i^\sigma \qquad [4.56]$$

These relations obtained from the second derivatives of the thermodynamic potential show that the direct and converse piezoelectric coefficients are equal (equation [4.54]). The same is applicable to thermal expansion coefficients and piezo-caloric coefficients (equation [4.55]) and also to pyroelectric coefficients and to electro-caloric coefficients (equation [4.56]).

The following relations can thus be determined:

$$\varepsilon_{ik} = S_{iklm}^{E,T} \sigma_{lm} + d_{lik}^T E_l + \alpha_{ik}^E \Delta T \qquad [4.57]$$

$$D_i = d_{ikl}^T \sigma_{kl} + \chi_{ik}^{\sigma,T} E_k + p_i^\sigma \Delta T \qquad [4.58]$$

$$\Delta S = \alpha_{ik}^E \sigma_{ik} + p_i^\sigma E_i + \frac{C^{\sigma,E}}{T} \Delta T \qquad [4.59]$$

The transformation from the tensor notation to the matrix notation can be done by contracting the indices and writing the constitutive equations in the following form:

$$\varepsilon = S^{E,T} \sigma + d^T E + \alpha^E \Delta T \qquad [4.60]$$

$$D = d^T \sigma + \chi^{\sigma,T} E + p^\sigma \Delta T \qquad [4.61]$$

$$\Delta S = \alpha^E \sigma + p^\sigma E + \frac{C^{\sigma,E}}{T} \Delta T \qquad [4.62]$$

If the temperature is assumed to be constant, the constitutive equations for the study of piezoelectric phenomena are written as follows:

$$\varepsilon = S^E \sigma + dE \tag{4.63}$$

$$D = d\sigma + \chi^\sigma E \tag{4.64}$$

$$\Delta S = \alpha^E \sigma + p^\sigma E \tag{4.65}$$

The constitutive equations for the couplings involving the magnetic field or other types of stresses can be established from potential functions obtained by the Legendre transform from the internal energy U in order to recover the relevant independent variables.

A magnetostrictive material changes its shape (crystalline structure) when it is exposed to a magnetic field. Magnetostrictive expansion can be used to create a linear motion with strong forces and very short response times. Magnetostrictive actuators are efficient, operate at low voltage, transmit large amounts of energy per volume and enable contactless action (distant magnetic field). The material can be deposited in thin films for the manufacturing of microsystems or nanosystems. An example of material is based on Terfenol, an alloy of Terbium, Dysprosium and iron ($Tb_{0.3}Dy_{0.7}Fe_{1.9}S$). It is a magnetostrictive material that makes it possible to gain an order of magnitude on the common piezoelectric components (known as giant magnetostriction). Another magnetostrictive material is Permendur (alloy composed of 49% Fe, 49% Co and 2% V).

Shape memory alloys (SMA) present phase transition phenomena, the alloy transforming from a given structure to another. The material changes its length as a function of temperature following a thermo-mechanical coupling. One of the most common alloys is a combination of nickel and titanium. This shape memory alloy can be processed so that it contracts when it reaches a defined temperature. When it cools, it returns to its original shape. This technology can be used to open a valve in a coffee machine at a defined temperature. When embedded in a deformable material in the form of wires, such materials enable a response depending on a current flow. The material retracts and can be used in robotics technology for feel-touch operation. Figure 4.10 shows the different steps in the development of such a material.

The metallurgical phases involved in the process are the martensite phase at low temperature and the austenite phase at high temperature. When there is a phase transformation, the SMA modifies its shape, hence it undergoes a two-way shape "memory" effect. The SMA process starts with the material annealed at high temperature in the austenite phase (Figure 4.10). The shape is in this way fixed in the material. During cooling, the material transforms into the martensite phase and adopts an interlaced crystalline structure. When subjected to a mechanical strain, the interlaced crystalline structure transforms to an asymmetric crystalline structure.

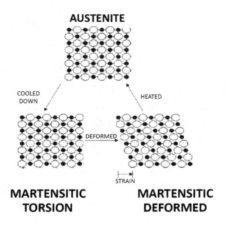

Figure 4.10. *Shape memory alloy development process*

During heating, the martensite phase transforms into austenite and the shape initially imposed by annealing is recovered. In this way, the permanent deformation created by the onset of the martensite phase is suppressed and the material returns to its initial state memorized during the annealing at austenitic temperatures. If the recovery is mechanically prevented, stresses up to 700 MPa can be developed.

Table 4.7 summarizes several orders of magnitude for several active materials that can be included in smart systems.

	Shape Memory Alloy	Piezoelectric	Giant magnetostrictive	Bimorph actuators
Physical phenomenon	Martensitic transformation	Piezoelectricity	Magnetostriction	Differential thermal expansion
External physical stress	Thermal	Electric field	Magnetic field	Thermal
Energy density (J/m^3)	10^6 to 10^7	10^2 (PZT) 10^3 (PMN)	10^6 to 10^7 (Terfenol D)	10^6 (Ni/Si)
Bandwidth	Low	High	High	Low
Working mode	Bending, torsion, tension, compression	Depends on the direction of the electric field	Depends on the direction of the magnetic field	Bending
Typical strain	**1% to 8%	0.12% to 0.15%	0.58% to 0.81%	5% to 23x10^{-4}
	*Strongly dependent on shape and size **Depends on lifetime specification (Maximal strain up to 15% of a monocrystal)			*Strongly dependent on shape and size

Table 4.7. *Orders of magnitude of the characteristics of some active materials*

4.4. Exercises on the application of active materials

4.4.1. *Strain tensor for 2D thin films*

In this exercise, the elements of a strain tensor are calculated for a plate or thin film, whose thickness is negligible compared to its lateral dimensions. The results can be applied to the calculation of the strain of a substrate subjected to a given temperature during thin film deposition, for example.

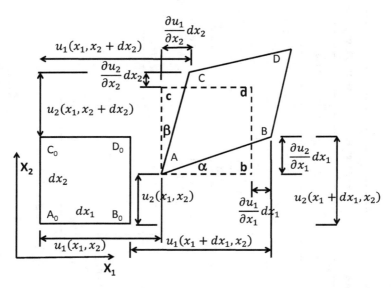

Figure 4.11. *Displacement vectors to calculate the strain tensor in 2D*

4.4.1.1. Questions

1) Recall the principle of the model used for the calculation of the strain tensor using Figure 4.11 as a reference.

2) Provide an interpretation of the square in Figure 4.11 in the model used for the calculation of the strain tensor using Figure 4.6 as a reference.

3) Provide an interpretation of the rhomboid in Figure 4.11 in the model used for the calculation of the strain tensor using Figure 4.6 as a reference.

Smart Materials   185

4) What does the vector A_0A represent?

5) What does the vector B_0B represent?

6) What does the vector C_0C represent?

Using the data in Figure 4.11, calculate:

7) $\tan \alpha$.

8) $\tan \beta$.

9) $u_1(x_1 + d\,x_1, x_2) - u_1(x_1, x_2)$ and give an interpretation of this result.

10) $u_2(x_1 + d\,x_1, x_2) - u_2(x_1, x_2)$ and give an interpretation of this result.

11) $u_1(x_1, x_2 + d\,x_2) - u_1(x_1, x_2)$ and give an interpretation of this result.

12) $u_2(x_1, x_2 + d\,x_2) - u_2(x_1, x_2)$ and give an interpretation of this result.

13) Calculate the components of the tensor based on elongations per unit length. Give the expressions of this tensor denoted by e_{ij}.

14) In the case of an overall rotation of the plane with an angle δ without strain, calculate the components of the tensor.

15) Prove that the second rank tensor T_{ij} can be written in the form of a symmetric tensor S_{ij} and of an antisymmetric tensor A_{ij}.

16) Give the expressions of S_{ij} and A_{ij} in the case of tensor e_{ij}.

17) Provide an interpretation of the terms S_{ij} and A_{ij}.

4.4.1.1. Solutions

1) Recall the principle of the model used for the calculation of the strain tensor using Figure 4.11 as a reference.

Solutions:

The hypothesis of the continuous medium is generally adopted in order to describe the strain of a material. At a point of the material localized by a

186   Applications and Metrology at Nanometer Scale 1

vector **r**, which in 2D has the components x_1 and x_2 and located at point A in the figure, consider small displacements represented by a translation vector **u(r)**, which is represented by vector A_0A in the figure.

During the deformation of the material, the variations of $u_1(x_1, x_2)$ along x_1 and x_2 are evaluated, as the quantities $du_1(x_1, x_2) = (\partial u_1/\partial x_1)dx_1 + (\partial u_1/\partial x_2)dx_2$ and $du_2(x_1, x_2) = (\partial u_2/\partial x_1)dx_1 + (\partial u_2/\partial x_2)dx_2$, which corresponds to four possible components for the mathematical representation of this deformation.

Taking Figure 4.6 of section 4.3.1 as a reference, the square denoted by ABCD is represented in Figure 4.11 as the unit area element in 2D considered at the point of coordinates x,y of edges dx and dy.

2) Provide an interpretation of the square in Figure 4.11 in the model used for the calculation of the strain tensor using Figure 4.6 as a reference.

Solutions:

A unit area element around point A_0 is considered in order to evaluate the strains of the material by determining how the vector **u(r)** is modified during a deformation. This is calculated from the derivatives of this vector with respect to coordinates by considering its modification at neighboring points B_0 and C_0 separated from A_0 by the distances dx_1 and dx_2, respectively. The square corresponds to the infinitesimal area in the surroundings of A_0 with its origin at A_0.

3) Provide an interpretation of the rhomboid in Figure 4.11 in the model used for the calculation of the strain tensor using Figure 4.6 as a reference.

Solutions:

When the material is deformed, at the point localized by **r** of components x_1, x_2 at A_0, and in its surroundings (B_0, C_0 and D_0), each component of **u(r)**, $u_1(x_1, x_2)$ and $u_2(x_1, x_2)$ is *a priori* subjected to different modifications depending on the position $x'_1 = x_1 + \alpha dx_1$ and $x'_2 = x_2 + \beta dx_2$ ($\alpha = 0$ or 1, $\beta = 0$ or 1). This different variation depending on the initial position A_0, B_0, C_0 or D_0, during the deformation, is represented on the figure by the vectors B_0B, C_0C and D_0D. The rhomboid is a schematic representation of the unit area deformed around point A separated from A_0 by $u(r)$.

4) What does the vector A_0A represent?

Solutions:

The vector A_0A in the figure is a translation vector $u(r)$ with its origin at A_0. Its components are $u_1(x_1, x_2)$ and $u_2(x_1, x_2)$.

5) What does the vector B_0B represent?

Solutions:

The vector B_0B in the figure is a translation vector $u(r+dr)$ with its origin at B_0. Its components are $u_1(x_1+d x_1, x_2)$ and $u_2(x_1+d x_1, x_2)$.

6) What does the vector C_0C represent?

Solutions:

The vector C_0C in the figure is a translation vector $u(r+dr)$ with its origin at C_0. Its components are $u_1(x_1, x_2+d x_2)$ and $u_2(x_1, x_2+d x_2)$.

7) Using the data in the figure, calculate $\tan \alpha$.

Solutions:

According to the figure, $\tan \alpha = du_2(x_1, x_2)/(d x_1+du_1(x_1, x_2))$. Since $du_1(x_1, x_2)$ is small compared to $d x_1$ and for small angles, then $\alpha = du_2(x_1, x_2)/dx_1 = \partial u_2/\partial x_1$.

8) Using the data in the figure, calculate $\tan \beta$.

Solutions:

According to the figure, $\tan \beta = du_1(x_1, x_2) / (d x_2+du_2(x_1, x_2))$. Since $du_2(x_1, x_2)$ is small compared to $d x_2$ and for small angles, then $\beta = du_1(x_1, x_2)/dx_2 = \partial u_1/\partial x_2$.

9) $u_1(x_1+d x_1, x_2) - u_1(x_1, x_2)$ and give an interpretation of this result.

Solutions:

$$u_1(x_1+d x_1, x_2) - u_1(x_1, x_2) = du_1(x_1, x_2) = (\partial u_1/\partial x_1)dx_1.$$

188 Applications and Metrology at Nanometer Scale 1

This difference represents the elongation of the component u_1 of vector u parallel to OX_1. The elongation per unit length is the strain along OX_1.

10) $u_2(x_1+d\,x_1, x_2) - u_2(x_1, x_2)$ and give an interpretation of this result.

Solutions:

$$u_2(x_1+d\,x_1, x_2) - u_2(x_1, x_2) = du_2(x_1, x_2) = (\partial u_2/\partial x_1)dx_1.$$

This difference represents the elongation of the component u_2 of the vector u by rotation through an angle α. The elongation per unit length is the angle of rotation on the side dx_1.

11) $u_1(x_1, x_2+d\,x_2) - u_1(x_1, x_2)$ and give an interpretation of this result.

Solutions:

$$u_1(x_1, x_2+d\,x_2) - u_1(x_1, x_2) = du_1(x_1, x_2) = (\partial u_1/\partial x_2)dx_2.$$

This difference represents the elongation of the component u_1 of vector u by rotation with an angle β. The elongation per unit length is the angle of rotation on the side dx_2.

12) $u_2(x_1, x_2+d\,x_2) - u_2(x_1, x_2)$ and give an interpretation of this result.

Solutions:

$$u_2(x_1, x_2+d\,x_2) - u_2(x_1, x_2) = du_2(x_1, x_2) = (\partial u_2/\partial x_2)dx_2.$$

This difference represents the elongation of the component u_2 of vector u parallel to OX_2. The elongation per unit length is the strain along OX_2.

13) Calculate the components of the tensor based on elongations per unit length. Give the expressions of this tensor denoted by e_{ij}.

Solutions:

Given the calculations from 9 to 12, the following relations can be written:

$$(\partial u_1/\partial x_1) = e_{11} \text{ and } (\partial u_2/\partial x_2) = e_{22}.$$

Similarly: $(\partial u_1/\partial x_2) = e_{12}$ and $(\partial u_2/\partial x_1)=e_{21}$.

14) In the case of an overall rotation of the plane with an angle δ without strain, calculate the components of the tensor.

Solutions:

$e_{11}=e_{22}=0$ and $e_{12} = - e_{21} = \delta$ (as $\alpha >0$, but $\beta<0$).

15) Prove that the second rank tensor T_{ij} can be written in the form of a symmetric tensor S_{ij} and of an antisymmetric tensor A_{ij}.

Solutions:

Given $S_{ij}= (1/2(T_{ij}+T_{ji})$; it can be verified that $S_{ji} = S_{ij}$.

Given $A_{ij}= (1/2(T_{ij}-T_{ji})$; it can be verified that $A_{ji} = -A_{ij}$.

where $T_{ij} = S_{ij} +A_{ij}$.

16) Give the expressions of S_{ij} and A_{ij} in the case of tensor e_{ij}.

Solutions:

$S_{ij}= (1/2(e_{ij}+e_{ji})$ and $A_{ij}= (1/2(e_{ij}-e_{ji})$.

17) Provide an interpretation of the terms of S_{ij} and A_{ij}.

Solutions:

For an overall rotation: $S_{11}=S_{22}=0$ and $S_{12}= S_{21}=0$.

$A_{11}=A_{22}=0$ and $A_{12}= e_{12}$ and $A_{21}= -e_{21}$ (axial vector equivalent to a second rank antisymmetric tensor: a rotation axis and an angle about this axis).

In this case, the symmetric tensor S_{ij} represents the strain:

$S_{12}= (1/2(e_{12}+e_{21})$, $S_{21}= (1/2(e_{21}+e_{12})$ and $S_{11}= e_{11}$ and $S_{22}= e_{22}$.

As shown in questions (9) and (12), S_{11} and S_{22} represent an elongation strain or an expansion.

In this case, S_{12} and S_{21} represent a shear or a sliding strain (except for block rotation).

4.4.2. A piezoelectric accelerometer

In a piezoelectric accelerometer, a case (see Figure 4.12) is supported by a piezoelectric material that supplies an electric charge Q proportional to the corrective force, hence to the mass displacement. An accelerometer in longitudinal compression mode is considered, as illustrated in Figure 4.12. When the case is subjected to an acceleration a, a force of inertia F compresses the piezoelectric material in parallel to the direction of the acceleration which is labeled as axis 3. The dimensions at rest of a piezoelectric layer are t for the thickness, L for the length and W for the width. Its capacitance is denoted by C, and the corresponding dielectric permittivity is ε_{33}. The objective is to determine the static characteristics of the accelerometer.

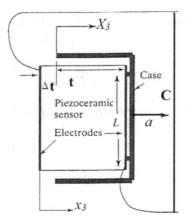

Figure 4.12. *Diagram of a piezoelectric accelerometer*

4.4.2.1. Questions

1) Find the vector expression of the strain tensor S of the layer.

2) The polarization of the layer being along the axis of deformation, find the vector expression of the polarization vector D of the layer.

3) Find the vector expression of the stress tensor T of the layer.

4) Relying on equations [4.63] and [4.64] expressed as $S = sT + dE$ and $D = dT + \varepsilon E$, deduce the equations verified by the piezoelectric layer for its mechanical strain and its electric polarization. (It is worth noting that there is no external electric stress, which means that $E = 0$.).

5) Based on the equations of question 4) show that: $\dfrac{Q}{\Delta t/t} = \dfrac{d_{33}LW}{s_{33}^{E}}$

6) Given: $t = 0.5$ mm; $L = 38.1$ mm; $W = 12.7$ mm; $\Delta t/t = 10^{-6}$; $d_{33} = 298.8 \ 10^{-12}$ CN-1; $s_{33}^{E} = 12.5 \ 10$ -12 m^2N^{-1}; $\varepsilon_{33} = 11.95 \ 10^{-9}$ Fm^{-1}. Calculate: the charge Q, the capacitance C of the layer and the voltage V across the layer.

4.4.2.2. Solutions

1) Find the vector expression of the strain tensor S of the layer.

$$S = \begin{pmatrix} S_1 = & 0 \\ S_2 = & 0 \\ S_3 = & \Delta t/t \\ S_4 = & \\ S_5 = & \\ S_6 = & \end{pmatrix}$$

2) The polarization of the layer being along the axis of deformation, find the vector expression of the polarization vector D of the layer.

$$D = \begin{pmatrix} D_1 = & 0 \\ D_2 = & 0 \\ D_3 = & Q/LW \end{pmatrix}$$

3) Find the vector expression of the stress tensor T of the layer.

$$T = \begin{pmatrix} T_1 = & 0 \\ T_2 = & 0 \\ T_3 = & F/LW \\ T_4 = & 0 \\ T_5 = & 0 \\ T_6 = & 0 \end{pmatrix}$$

4) Relying on equations [4.63] and [4.64] expressed as $S = sT+dE$ And $D=dT+\varepsilon E$, deduce the equations verified by the piezoelectric layer for its mechanical strain and its electric polarization. (It is worth noting there is no external electric stress, which means that $E = 0$.)

$$1)\ S_3 = s_{33}^E T_3 \Rightarrow \frac{\Delta t}{t} = s_{33}^E \frac{F}{LW} \text{ and } 2)\ D_3 = d_{33}T_3 \Rightarrow \frac{Q}{LW} = d_{33}\frac{F}{LW}$$

5) Based on the equations of question 4 show that: $\dfrac{Q}{\Delta t/t} = \dfrac{d_{33}LW}{s_{33}^E}$

Equations 1 and 2 from the previous solution lead to:

$$\frac{Q}{LW}\bigg/\frac{\Delta t}{t} = \frac{d_{33}\dfrac{F}{LW}}{s_{33}^E\dfrac{F}{LW}} \Rightarrow \frac{Q}{\Delta t}\bigg/\frac{}{t} = \frac{d_{33}LW}{s_{33}^E}$$

6) Given: $t = 0.5$ mm; $L = 38.1$ mm; $W = 12.7$ mm; $\Delta t/t = 10^{-6}$; $d_{33} = 298.8\ 10^{-12}\ CN^{-1}$; $s^E_{33} = 12.5\ 10^{-12}\ m^2N^{-1}$; $\varepsilon_{33} = 11.95\ 10^{-9}\ Fm^{-1}$. Calculate: the charge Q, the capacitance C of the layer and the voltage V across the layer.

$$Q = \frac{\Delta t}{t}\bigg/\frac{d_{33}LW}{s_{33}^E} \qquad \text{AN}: Q = 1{,}1157\ 10^{-8}\ C$$

$$C = \frac{\varepsilon \times \text{surface}}{\text{distance}} \Rightarrow C = \frac{\varepsilon_{33} \times LW}{t} \qquad \text{A.N. } C = 1{,}156\ 10^{-8}\ F$$

$$Q = CV \Rightarrow V = Q/C = 1\ \text{V}.$$

4.4.3. Piezoelectric transducer

Piezoelectric materials can be stacked in the form of layers either mechanically in series or electrically in parallel as shown in Figure 4.13.

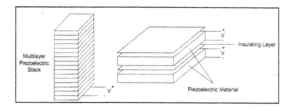

Figure 4.13. *Diagram of a piezoelectric transducer*

The stack on the right of this figure is considered for this exercise. Equations [4.63] and [4.64] are expressed as $S=sT+dE$ and $D=dT+\varepsilon E$, respectively. The objective is to determine the force, the displacement and the capacitance as a function of external excitations in the absence of stress and strain. The following conventions are adopted for the usual axes Ox, Oy and Oz by denoting them as 1, 2 and 3, respectively, as shown in Figure 4.14 and the stresses and strains are all considered zero, except for S_3 and T_3, respectively. Similarly, the electric field and the electric displacement are parallel to axis 3, which means that $E_1 = E_2 = 0$ and $D_1 = D_2 = 0$.

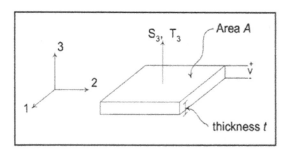

Figure 4.14. *Piezoelectric layer*

4.4.3.1. Questions

1) Express S_3 and D_3 as a function of T_3 and E_3. $x_i = \int_0^t S_3 dx_3$

194 Applications and Metrology at Nanometer Scale 1

2) The elongation of the ith layer is given by:

Deduce x_i as a function of s_{33}, d_{33}, T_3 and E_3.

3) Express E_3 and T_3 as a function of the given data.

4) Deduce from question 3, the final expression of x_i.

5) Assume that the transducer is composed of n layers and that the total length of the transducer is L. What is its total elongation X in this case?

6) The amount of charges due to the ith layer is given by:

$$q_i = \iint_{surface} D_3 dx_1 dx_2$$

Deduce from this expression, q_i as a function of d_{33}, ε_{33}, T_3 and E_3. Then express the total charge Q on the faces of the transducer as a function of d_{33}, L, t, F, n, A, V and ε_{33}.

7) Express the relation between X and Q and F and V in matrix form. Give an interpretation of the various elements of this 2 x 2 matrix.

8) Consider the situation of a zero force or of a zero strain on the transducer as shown in Figure 4.15(a) and (b).

Find in each case the expression of the force and of the displacement of the transducer and of the charge.

9) Deduce from results of question 8) that the capacitance C can be expressed in the following form: $(1 - k_{33}^2) (n\varepsilon_{33}A) /t$. Find in each case the expression of k_{33}^2.

10) Calculate the stress for an electric field of 0.5 MV/m and a strain of 10 MPa applied to the piezoelectric material. The inverse of Young's modulus is, $1/Y = s = 20 \; 10^{-12} \; m^2/N$ and the piezoelectric stress coefficient is $d_{33} = 650 \; 10^{-12} \; m/V$.

11) Find the geometry of a stack of piezoelectric layers subjected to a blocking force of 1000 N and a free displacement of 30 μm. The inverse of Young's modulus is, $1/Y = s = 20 \; 10^{-12} \; m^2/N$, the piezoelectric stress

coefficient is $d_{33} = 650 \; 10^{-12}$ m/V, the layer thickness is 254 μm, and the maximum allowed electric field is 0.3 MV/m.

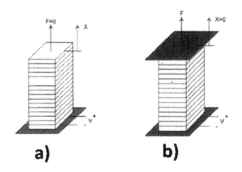

Figure 4.15. *a) Zero forces and b) zero strain*

4.4.3.2. *Solutions*

1) Express S_3 and D_3 as a function of T_3 and E_3.

From

$$\underline{S} = s\underline{T} + d\underline{E}$$
$$\underline{D} = d\underline{T} + \varepsilon \underline{E}$$

It can be deduced that:

$$S_3 = S_{33}^E T_3 + d_{33} E \text{ and } D_3 + d_{33} T_3 + \varepsilon_{33}^T E$$

2) The elongation of the ith layer is given by: $x_i = \int_0^t S_3 dx_3$. Deduce x_i as a function of s_{33}, d_{33}, T_3 and E_3.

For the ith layer: $x_i = \int_0^t (s_{33}^E T_3 + d_{33} E_3) dx_3 = (s_{33}^E T_3 + d_{33} E_3) t$

3) Express E_3 and T_3 as a function of the given data:

$$E_3 = \frac{V}{t} \text{ and } T_3 = \frac{F}{A}$$

If there are no stress and strain, then:

196 Applications and Metrology at Nanometer Scale 1

4) Deduce the final expression of x_i

Result:

$$x_i = s_{33}^E \frac{F}{A} t + d_{33} \frac{V}{t} t = s_{33}^E \frac{F}{A} t + d_{33} V$$

5) Assume that the transducer is composed of n layers and that the total length of the transducer is L. What is its total elongation X in this case?

Result:

$$X = \sum_i x_i = nx_i = \frac{L}{t}(s_{33}^E \frac{F}{A} t + d_{33}V) = s_{33}^E \frac{FL}{A} + d_{33}V \frac{L}{t}$$

6) The amount of charges due to the ith layer is given by:

$$q_i = \iint_{surface} D_3 dx_1 dx_2$$

Deduce from this expression, q_i as a function of d_{33}, ε_{33}, T_3 and E_3. Then express the total charge Q on the faces of the transducer as a function of d_{33}, L, t, F, n, A, V and ε_{33}.

Result:

$$q_i = \iint_S (d_{33}T_3 + \varepsilon_{33}^T E_3) dx_1 dx_2 = d_{33}T_3 A + \varepsilon_{33}^T E_3 A$$

Hence:

$$Q = \sum_i q_i = nq_i = \frac{L}{t} d_{33} \frac{F}{A} A + \frac{L}{t} \varepsilon_{33}^T \frac{V}{t} A = \frac{L}{t} d_{33} F + \varepsilon_{33}^T n \frac{V}{t} A$$

7) Express the relation between X and Q and F and V in matrix form. Give an interpretation of the various elements of this 2 x 2 matrix.

$$X = \frac{S_{33}^E L}{A} F + \frac{d_{33}L}{t} V \text{ and } Q = \frac{d_{33}L}{t} F + \frac{\varepsilon_{33}^T nA}{t} V$$

It can be written that:

Hence, the matrix form:
$$\begin{pmatrix} X \\ Q \end{pmatrix} = \begin{pmatrix} \dfrac{s_{33}^E L}{A} & \dfrac{d_{33}L}{t} \\ \dfrac{d_{33}L}{t} & \dfrac{\varepsilon_{33}^T nA}{t} \end{pmatrix} \begin{pmatrix} F \\ V \end{pmatrix}$$

The various terms can be interpreted as follows:

The first term (M_{11}) is the displacement under the effect of the mechanical force in the absence of electric stress (or electric field): it is the compliance or the mechanical elasticity coefficient or the inverse Young's modulus ($X=CF$ or $F=YX$)) (m/N).

The second term (M_{12}) is the displacement under the effect of the electric stress (or electric field): it is the piezoelectric coupling coefficient (m/V).

The third term (M_{21}) corresponds to the appearance of electric charges under the effect of a mechanical force in the absence of electric stress (or electric field): it is the piezoelectric coupling coefficient (m/V).

The fourth term (M_{22}) is the displacement under the effect of electric constraint (or electric field): it is the capacitance ($Q=CV$; C in Farads).

8) Consider the situation of a zero force or of a zero strain on the transducer as shown in Figure 4.15(a) and (b).

Find in each case the expression of the force and of the displacement of the transducer and of the charge.

a) Zero force: $F= X = \dfrac{d_{33}L}{t}V = nd_{33}V$ and $\quad Q = \dfrac{\varepsilon_{33}^T nA}{t}V = C^T V$

b) Zero strain, X=0: $0 = \dfrac{s_{33}^E L}{A}F + \dfrac{d_{33}L}{t}V \Rightarrow F = -\dfrac{d_{33}A}{s_{33}^E t}V$

and $Q = \dfrac{d_{33}L}{t}F + \dfrac{\varepsilon_{33}^T nA}{t}V = -\dfrac{d_{33}^2 AL}{s_{33}^E t^2}V + \dfrac{\varepsilon_{33}^T nA}{t}V = (-\dfrac{d_{33}^2}{s_{33}^E} + \varepsilon_{33}^T)\dfrac{nA}{t}V$

9) Deduce from the results of question 8 that the capacitance C can be expressed in the following form: $(1 - k_{33}^2)$ $(n\varepsilon_{33}A)$ $/t$. Find in each case the expression of k_{33}^2.

198 Applications and Metrology at Nanometer Scale 1

a) $k_{33}^2 = 0$

b) $Q = CV \Rightarrow C = (-\dfrac{d_{33}^2}{s_{33}^E} + \varepsilon_{33}^T)\dfrac{nA}{t} = (1 - \dfrac{d_{33}^2}{\varepsilon_{33}^T s_{33}^E})\dfrac{nA\varepsilon_{33}^T}{t} = (1 - k_{33}^2)C^T$

10) Calculate the stress for an electric field of 0.5 MV/m and a strain of 10 MPa applied to the piezoelectric material. The inverse of Young's modulus is, $1/Y = s = 20$ x 10^{-12} m^2/N, and the piezoelectric stress coefficient is $d_{33} = 650$ x 10^{-12} m/V.

$$\frac{X}{L} = s_{33}^E \frac{F}{A} + d_{33}\frac{V}{t} \Rightarrow S = s_{33}^E T + d_{33}E$$

Hence: $S = (20 \ 10^{-12} \text{ x } 10 \ 10^6) + (650 \ 10^{-12} \text{ x } 0.5 \ 10^6) = 525$ x 10^{-6}.

11) Find the geometry of a stack of piezoelectric layers subjected to a blocking force of 1000 N and a free displacement of 30 µm. The inverse of Young's modulus is, $1/Y = s = 20$ x 10^{-12} m^2/N, the piezoelectric stress coefficient is $d_{33} = 650$ x 10^{-12} m/V, the layer thickness is 254 µm, and the maximum allowed electric field is 0.3 MV/m.

The free displacement is: $\dfrac{X}{L} = d_{33}\dfrac{V}{t} \Rightarrow L = \dfrac{X}{d_{33}E}$

Hence: $L = (30$ x $10^{-6} / (650$ x 10^{-12} x 0.3 x $10^6) = 0.154$ m.

The blocking force is: $F = -\dfrac{d_{33}A}{s_{33}^E t}V = -\dfrac{d_{33}A}{s_{33}^E}E \Rightarrow A = \left|\dfrac{Fs_{33}^E}{d_{33}E}\right|$

Hence: $A = (1000$ x 20 x $10^{-12)} / (650 \ 10^{-12}$ x 0.3 x $10^6) = 1.03$ x 10^{-4} m^2.

The calculated number of layers is: $n = L/t = (0.154/254$ x $10^{-6}) = 607$.

4.4.4. Piezoelectric sensor

Piezoelectric materials can be stacked as double layers (see Figure 4.16(a)) that constitute a force sensor as in a scanning probe microscope used for studying the topography of nanomaterial surfaces.

The objective is to determine the force, the displacement and the capacitance as a function of external excitations in the absence of stress and strain. Equations [4.63] and [4.64] are expressed as $S=sT+dE$ and $D=dT+\varepsilon E$, respectively. By convention, the Ox, Oy and Oz axes are denoted by 1, 2 and

3, respectively, as shown in Figure 4.16(b), and all the stresses and strains are considered to be zero, except for S_1 and T_1, respectively. Similarly, the electric field and the electric displacement are parallel to axis 3, meaning that $E_1 = E_2 = 0$ and $D_1 = D_2 = 0$.

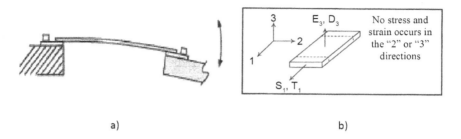

a) b)

Figure 4.16. a) Piezoelectric sensor and b) piezoelectric layer

4.4.4.1. Questions

1) Based on the following general constitutive equations of a piezoelectric material: $\begin{array}{l} \underline{S} = s\underline{T} + d\underline{E} \\ \underline{D} = d\underline{T} + \varepsilon \underline{E} \end{array}$ express S_1 and D_3 as a function of T_1 and E_3.

Consider the stack shown in Figure 4.17 and the associated electric circuit such that the top and bottom layers are subjected to an electric field whose direction is, respectively, parallel and antiparallel to the direction of the layer polarization. It is assumed that $L \gg t$.

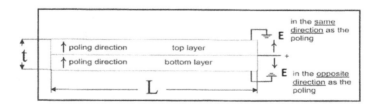

Figure 4.17. Piezoelectric double layer

2) Consider that the stress T is zero. Calculate S_1 for each layer. Compare the sign and comment the diagrams in Figure 4.18(a) and (b).

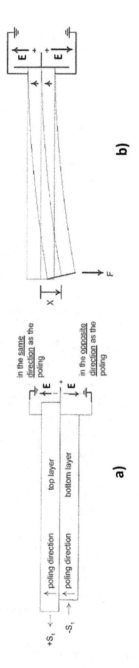

Figure 4.18. a) *Piezoelectric double layer and* b) *bending of layers*

The bending of the layer is parameterized by the X variable, which gives the vertical displacement of the contact plane between the two layers.

Given: $X = 4\dfrac{s_{11}^E L^3}{wt^3} F + 3\dfrac{d_{31} L^2}{t^2} V$

3) What is the free displacement X of the two layers ($F=0$)?

4) Calculate the force F required to cancel the displacement X ($X=0$).

5) What is the effect of the ratio L/t on X and F calculated in solutions 3 and 4?

4.4.4.2. Solutions

1) Based on the following general constitutive equations of a piezoelectric material: $\begin{aligned} \underline{S} &= s\underline{T} + d\underline{E} \\ \underline{D} &= d\underline{T} + \varepsilon\underline{E} \end{aligned}$ express S_1 and D_3 as a function of T_1 and E_3.

Based on the general equation and since only T_1, S_1, E_3 and D_3 are different from zero, the relations can be written as follows:

$$S_1 = s_{11}^E T_1 + d_{13} E_3$$
$$D_3 = d_{31} T_1 + \varepsilon_{33}^T E_3$$

2) Consider that the stress T is zero. Calculate S_1 for each layer. Compare the sign and comment the diagrams in Figure 4.18(a) and (b).

In the absence of stress T, equation 1 of the first question, for the top and bottom layers, leads, respectively, to:

$$S_1 = +d_{13} E_3$$
$$S_1 = -d_{13} E_3$$

202 Applications and Metrology at Nanometer Scale 1

Comparison: the signs are opposite, as the electric potential difference is in opposition.

Comments: an expansion of the top layer and a contraction of the bottom layer can be noted in Figure 4.18(a), which is in agreement with the results of the calculations. The bending of the double layer minimizes the stresses and prevents the fracture of the two layers.

3) What is the free displacement X of the two layers ($F=0$)?

The calculation leads to:

$$X = 3\frac{d_{31}L^2}{t^2}V$$

4) Calculate the force F required for cancelling out the displacement X ($X=0$).

The calculation yields:

$$F = -\frac{(3\frac{d_{31}L^2}{t^2}V)}{(4\frac{s_{11}^E L^3}{wt^3})} = -(\frac{d_{31}wt}{s_{11}^E L})V$$

5) What is the effect of the ratio L/t on X and F calculated in solutions 3 and 4?

X depends on L/t while F is inversely proportional to L/t. Increasing L/t leads to an increase of X and a decrease of the force, F.

4.5. Appendix: crystal symmetry

In condensed phase, only the rotation axes of orders 1, 2, 3, 4 and 6 enable a periodic pavement of the space compatible with crystalline lattices. The symmetry elements used by the crystallographers to define the symmetry about a point in space, for example, the central point of the unit cell, are: a point (center) of symmetry, rotation axes, mirror planes, and combinations thereof. Using these elements of symmetry, it is shown in crystallography that there are 32 symmetry point groups (Table 4.8) based on symmetry operations. The various symmetry elements are divided into

Smart Materials 203

seven crystalline systems as follows: triclinic, monoclinic, orthorhombic, tetragonal, rhombohedral, hexagonal and cubic.

Crystalline system	Number of point groups	Herman-Mauguin	Schoenflies
Triclinic	2	$1, \bar{1}$	C_1, C_i
Monoclinic	3	$2, m, 2/m$	C_2, C_s, C_{2h}
Orthorhombic	3	$222, 2mm, mmm$	$D_2, C_{2v}, D_{2h},$
Rhombohedral	5	$3, 32, 3m$ $\bar{3}, \bar{3}m$	C_3, D_3, C_{3v} S_6, D_{3d}
Hexagonal	7	$6, 622, 6mm, 6/m,$ $\bar{6}, \bar{6}2m, 6/mmm$	C_6, D_6, C_{6v}, C_{6h} C_{3h}, D_{3h}, D_{6h}
Tetragonal (quadratic)	7	$4, \bar{4}, 422, 4mm,$ $\bar{4}2m, 4/m, 4/mmm$	$C_4, S_4, D_4, C_{4v},$ D_{2d}, C_{4h}, D_{4h}
Cubic	5	$23, m3, 432,$ $\bar{4}3m, m3m$	T, T_h, O T_d, O_h

Table 4.8. *The point groups of 32 classes of symmetry and the equivalence between Schoenflies and Hermann–Mauguin symbols*

There are special point groups that can be added to the above groups. Among them, the continuous point groups have an infinite number of symmetry elements, the axial and spherical groups of symmetry used for listing the groups characterized by the presence of several rotational symmetry axes of high order n associated with regular polyhedrons, Platonic solids, such as the triangular tetrahedron, the cube with six square faces, the octahedron with eight triangular faces, the dodecahedron with 12 regular pentagonal faces with three pentagonal faces joined at one point and the icosahedron with 20 equilateral triangular faces with five faces joined at one point.

Table 4.9 summarizes the two conventions used to represent a symmetry element in a crystal.

Symmetry operation	Herman-Mauguin	Schoenflies
Identity	1	C_1 or E
Rotation $(2\pi/n)$	n	C_n
Mirror	m	σ, C_s, S_1
Inversion	$\bar{1}$	i, C_i, S_2
Rotation-Reflection	-	S_{2n}
Rotation-Inverison	$\bar{n}$	-

Table 4.9. *Schoenflies and Hermann–Mauguin notation*

The convention used in condensed phase is that of Hermann–Mauguin, which is preferred in crystallography to that of Schoenflies which is used in the spectroscopy study of atoms and molecules: the cyclic groups are denoted by their order n (rotational axis of symmetry of order n), a plane of symmetry is denoted by m and the group is denoted by nm ($C_{2v} \equiv 2m$); if the plane is perpendicular to the axis of symmetry, n/m ($C_{2h} \equiv 2/m$) is used; the presence of a center of symmetry is indicated by a bar above the order of the group ($C_i = S_2 \equiv \bar{1}$, $S_4 \equiv \bar{4}$, $S_6 = C_i \otimes C_3 \equiv \bar{3}$). For example, $O_h \equiv m3m$ and $D_{3h} \equiv \bar{3}m2$, etc.

Appendix

Propagation of a Light Ray

In light diffraction problems, the amplitude of a wave $E_z(x,y)$ on a surface located in a plane z is determined by the amplitude $E_0(x,y)$ of a wave in the plane $z = 0$.

In Chapter 3 of [DAH 16], equation 3.6 gives the expression of an electromagnetic wave traveling along Oz, in the form $u(z\text{-}vt)=acos(k(z\text{-}ct))$, where v is the speed of propagation of the wave in a medium of index n, a is its amplitude and $k=2\pi n/\lambda$. In a vacuum, $n=1$ and in a three-dimensional space the monochromatic plane wave of the angular frequency $\omega=kc$ and the wave vector $k = (k_x, k_y, k_z)$ can be expressed in the form $E = E_0 \exp i(2\pi/\lambda$ $(\alpha x+\beta y+\gamma z) - \omega t)$, where the components of the wave vector have the form: $k_x=2\pi\alpha/\lambda$, $k_y=2\pi\beta/\lambda$ and $k_x=2\pi\gamma/\lambda$, with α, β, γ being the direction cosines of the wave vector **k** and λ being the wavelength. This expression can be obtained by solving the Helmholtz equation (equation 3.7, [DAH 16]):

$$\Delta\vec{E} + \frac{\omega^2}{c^2}\vec{E} = \vec{0} \qquad\qquad [\text{A.1}]$$

where $\Delta\vec{E} = \vec{\nabla}^2\vec{E}$, using Green's function $G_k(\vec{r},\vec{r}_0)$ of the Helmholtz equation that verifies that:

$$(\nabla^2 + k^2)G_k(\vec{r},\vec{r}_0) = \delta(\vec{r} - \vec{r}_0) \qquad\qquad [\text{A.2}]$$

given that $G_k(\vec{r},\vec{r}_0) = -\frac{exp(ik|\vec{r}-\vec{r}_0|)}{4\pi|\vec{r}-\vec{r}_0|}$, where the unit vector is $\vec{e}_r = \frac{(\vec{r}-\vec{r}_0)}{|\vec{r}-\vec{r}_0|}$. The solution is readily obtained, since for the k mode of the electric field generated by a source placed at $\vec{r}_0$, $\vec{E}_k(\vec{r},\vec{r}_0) = -\frac{exp(ik|\vec{r}-\vec{r}_0|)}{4\pi|\vec{r}-\vec{r}_0|}\vec{e}_r$.

The propagation of a wave through obstacles (slits or opaque objects) can be determined using the Green–Ostrogradsky theorem and the Green function $G_k(\vec{r}, \vec{r}_0)$ of the Helmholtz equation.

Based on the relation:

$$\vec{\nabla}(u\vec{\nabla}v - v\vec{\nabla}u) = u\vec{\nabla}^2 v - v\vec{\nabla}^2 u = u\Delta v - v\Delta u \qquad [A.3]$$

where u and v are solutions to the Helmholtz equation and the Green–Ostrogradsky theorem, which enables a volume integral to be transformed into a surface integral:

$$\int (u\Delta v - v\Delta u)d\tau = \int (u\vec{\nabla}^2 v - v\vec{\nabla}^2 u)d\tau = \oint (u\vec{\nabla}v - v\vec{\nabla}u)d\vec{s} \qquad [A.4]$$

It can be written that:

$$\int \left(\vec{E}(\vec{r})\Delta G_k(\vec{r}, \vec{r}_0) - G_k(\vec{r}, \vec{r}_0)\Delta\vec{E}(\vec{r}) \right) d\tau = \int \vec{E}(\vec{r})\delta(\vec{r} - \vec{r}_0)d\tau$$

$$- \int \vec{E}(\vec{r})k^2 G_k(\vec{r}, \vec{r}_0)d\tau + \int G_k(\vec{r}, \vec{r}_0)k^2\vec{E}(\vec{r})d\tau = \vec{E}(\vec{r}_0) \qquad [A.5]$$

hence

$$\vec{E}(\vec{r}_0) = \oint \left(\vec{E}(\vec{r})\vec{\nabla}G_k(\vec{r}, \vec{r}_0) - G_k(\vec{r}, \vec{r}_0)\vec{\nabla}\vec{E}(\vec{r}) \right) d\vec{s}$$

$$= \oint (\vec{E}(\vec{r})\{\vec{\nabla}G_k(\vec{r}, \vec{r}_0)d\vec{s} - ikG_k(\vec{r}, \vec{r}_0)ds\}) \qquad [A.6]$$

where the surface integral surrounds the point located at $\vec{r}_0$ and $\vec{r}$ indicates the position of the surface element $d\vec{s}$, (Figure A.1) so that $\vec{\nabla}\vec{E}(\vec{r}).d\vec{s} = ik\vec{E}(\vec{r})ds$, for a wave that propagates in the volume surrounded by an integration surface ds. For a wave that propagates outside the surface, the relation is $\vec{\nabla}\vec{E}(\vec{r}).d\vec{s} = -ik\vec{E}(\vec{r})ds$.

Given that $G_k(\vec{r}, \vec{r}_0) = -\frac{exp(ik|\vec{r}-\vec{r}_0|)}{4\pi|\vec{r}-\vec{r}_0|}$, then:

$$\vec{\nabla}G_k(\vec{r}, \vec{r}_0) = ikG(\vec{r}, \vec{r}_0)\vec{e}_r + \frac{exp(ik|\vec{r} - \vec{r}_0|)}{4\pi|\vec{r} - \vec{r}_0|^2}\vec{e}_r$$

$$= \left(\frac{ik}{|\vec{r}-\vec{r}_0|} - \frac{1}{|\vec{r}-\vec{r}_0|^2} \right) (\vec{r} - \vec{r}_0)G_k(\vec{r}, \vec{r}_0) \qquad [A.7]$$

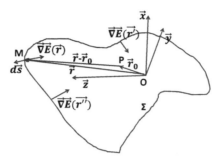

Figure A.1. *Kirchhoff's integral over a surface Σ passing through a point M surrounding point P. For a color version of this figure, see www.iste.co.uk/dahoo/metrology1.zip*

At a large distance compared to the wavelength such that $k|\vec{r}-\vec{r}_0| \gg 1$, the solution is:

$$\vec{E}(\vec{r}_0) = \oint \left(\vec{E}(\vec{r}) \left\{ \frac{ik}{|\vec{r}-\vec{r}_0|}(\vec{r}-\vec{r}_0)G_k(\vec{r},\vec{r}_0)d\vec{s} - ikG_k(\vec{r},\vec{r}_0)ds \right\} \right) \quad [\text{A.8}]$$

or:

$$\vec{E}(\vec{r}_0) = -\frac{ik}{4\pi} \oint \vec{E}(\vec{r}) \frac{exp(ik|\vec{r}-\vec{r}_0|)}{|\vec{r}-\vec{r}_0|^2} \left((\vec{r}-\vec{r}_0)d\vec{s} - |\vec{r}-\vec{r}_0|ds\right) \quad [\text{A.9}]$$

This solution corresponds to the formulation of the Huygens–Fresnel principle or the Huygens–Fresnel equation obtained using Kirchhoff's integral, which is the Fresnel–Kirchhoff diffraction formula.

When a monochromatic plane wave traveling in parallel to Oz meets a screen (E_0) with a slit (Figure A.2), a diffraction pattern appears in the observation plane. For the sake of simplicity, assume that the slit plane is perpendicular to the direction of propagation of the plane wave and that the amplitude and the gradient of the electric field of the wave are constant on the surface of the slit in the vicinity of M (Figure A.2) and zero everywhere else. The amplitude on the other side of the slit at a point P located at $\vec{r}_0$ is given by the Fresnel–Kirchhoff diffraction formula. If the slit is in the Oxy plane (Figure A.2), and the dimensions are small compared to the position of point P at $\vec{r}_0$, the amplitude of the field at P is given by:

$$\vec{E}(\vec{r}_0) = -\frac{ik\vec{E}(\vec{r})}{4\pi R}(cos\theta + 1) \oint exp(ik|\vec{r}-\vec{r}_0|)dxdy \quad [\text{A.10}]$$

where R is the distance between the central point of the slit and point P (in a first approximation, it is the same distance for all the points of the slit), and θ is the angle between the Oz axis and the vector $\overrightarrow{MP} = \vec{r}_0 - \vec{r}$. The origin of coordinates is considered in the plane containing the diffraction slit.

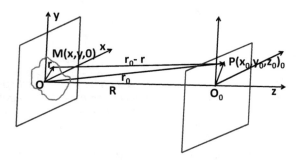

Figure A.2. *Diagram of diffraction in a plane containing P through a slit containing M. For a color version of this figure, see www.iste.co.uk/dahoo/metrology1.zip*

The module of the vector $\vec{r}_0 - \vec{r}$ is given by:

$$|\vec{r}_0 - \vec{r}| = \sqrt{r_0^2 - 2\vec{r}_0 \cdot \vec{r} + r^2}.$$

Considering the conditions of Fraunhofer diffraction in the far field, with an origin O in the slit plane (Figure A.2), where R is the distance between the observation plane and the diffraction plane, the following relation can be obtained:

$$exp(ik|\vec{r} - \vec{r}_0|) = exp\left(ik\sqrt{r_0^2 - 2\vec{r}_0 \cdot \vec{r} + r^2}\right) = e^{ikR} exp(-ik(\alpha x + \beta y)) \qquad [A.11]$$

where: $\alpha = \frac{x_0}{R}$ and $\beta = \frac{y_0}{R}$.

According to the Gaussian approximation of paraxial rays, the field at P is:

$$\vec{E}(\vec{r}_0) = A \oiint exp(-ik(\alpha x + \beta y)) dx dy \qquad [A.12]$$

Fraunhofer diffraction at large distance can be calculated using this formula. For a circular hole, the formula is:

$$\vec{E}(\vec{r}_0) = A \oiint exp(-ik(\alpha\rho cos\varphi))\rho d\rho d\varphi$$

$$= 2\pi A \int_0^a J_0(k\alpha\rho)\rho d\rho = 2\pi A a^2 \frac{J_1(k\alpha\rho)}{k\alpha\rho} \qquad [A.13]$$

where $J_0(x)$ and $J_1(x)$ are zero and the first-order Bessel functions.

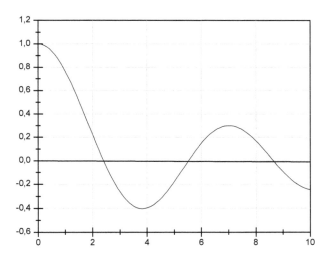

Figure A.3. *Distribution of the diffraction amplitude through a circular hole. For a color version of this figure, see www.iste.co.uk/dahoo/metrology1.zip*

Fresnel diffraction is obtained for an observation plane that is closer to the slit. In this case, second-order terms in x should be considered. Moreover, it is wise to use a coordinate system so as to get rid of the terms of order 1 in x and y. Assume that the slit is in the Oxy plane and the Oz axis is a line perpendicular to the plane that passes through the point of observation r₀. In this system of coordinates, $\alpha = \beta = \vec{r}.\vec{r}_0$=0. Under these conditions, the following expression is obtained:

$$exp(ik|\vec{r} - \vec{r}_0|) = exp\left(ik\sqrt{r_0^2 - 2\vec{r}_0.\vec{r} + r^2}\right) = e^{ikR} exp\left(\frac{ik(x^2+y^2)}{2R}\right) [A.14]$$

210   Applications and Metrology at Nanometer Scale 1

In the case of the Gaussian approximation of paraxial rays, the field at P is given by:

$$\vec{E}(\vec{r_0}) = A \int exp \left(\frac{ikx^2}{2R} \right) dx \int exp \left(\frac{iky^2}{2R} \right) dy \qquad [A.15]$$

which can be used to calculate the Fresnel diffraction pattern.

References

[ABB 73] ABBE E., "Beiträge zur theorie des mikroskops und der mikroskopischen wahrnehmung", *Arch. Mikroskop. Anat.*, vol. 9, pp. 413–418, 1873.

[ASH 72] ASH E.A., NICHOLLS G., "Super-resolution aperture scanning microscope", *Nature*, vol. 237, p. 510A, 1972.

[ASP 82] ASPECT A., DALIBARD J., ROGER G., "Experimental test of Bell's inequalities using time varying analyzers", *Phys. Rev. Lett.*, vol. 49, no. 25, pp. 1804–1807, 1982.

[BAR 48] BARDEEN J., BRATTAIN W.H., "The transistor, a semi-conductor triode", *Phys. Rev.*, no. 74, pp. 230–231, 1948.

[BET 86] BETZIG E., LEWIS A., HAROOTUNIAN A. *et al.*, "Near-field scanning optical microscopy (NSOM); development and biophysical applications", *Biophys. J.*, vol. 49, pp. 269–279, 1986.

[BET 87] BETZIG E., ISAACSON M., LEWIS A., "Collection mode near-field optical microscopy", *Appl. Phys. Lett.*, vol. 51, pp. 2088–2090, 1987.

[BIN 82a] BINNIG G., ROHER H., GERBER C. *et al.*, "Tunneling through a controllable vacuum gap", *Appl. Phys. Lett.*, vol. 40, pp. 178–180, 1982.

[BIN 82b] BINNIG G., ROHER H., "Scanning tunneling microscopy", *Helv. Phys. Acta*, vol. 55, pp. 726–735, 1982.

[BIN 83] BINNIG G., ROHER H., "Scanning tunneling microscopy", *Surf. Sci.*, vol. 126, pp. 236–244, 1983.

[BIN 86a] BINNIG G., ROHRER H. "Scanning tunneling microscopy", *IBM J. Res. Dev.*, vol. 30, pp. 355–369, 1986.

[BIN 86b] BINNIG G., QUATE C.F., GERBER C., "Atomic force microscope", *Phys. Rev. Lett.*, vol. 56, pp. 930–933, 1986.

[BIO 56] BIOT M.A., "Thermoelasticity and irreversible thermodynamics", *J. Appl. Phys.*, vol. 27, pp. 240–253, 1956.

[BRA 78] BRACEWELL R., *The Fourier Transform and its Applications*, McGraw-Hill Book Co., 1978.

[BRI 38] BRILLOUIN L., *Les tenseurs en mécanique et en élasticité*, Editions Masson et Cie, Paris, 1938.

[BRO 68] BROGLIE DE L., *Ondes électromagnétiques et photons*, Gauthier-Villars, Paris, 1968.

[BRU 55] BRUHAT G., *Mécanique*, Editions Masson et Cie., Paris, 1955.

[BRU 62] BRUHAT G., *Thermodynamique*, Editions Masson et Cie., Paris, 1962.

[BRU 65] BRUHAT G., *Cours de physique générale optique*, Editions Masson et Cie, Paris, 1965.

[COA 12] COANGA J.M., ALAYLI N., DAHOO P.R., "Metrological applications of ellipsometry with temperature change", *Proceedings of 4th International Metrology Conference 23–27 April 2012 (CAFMET 2012) Marrakech, Morocco*, Curran Associates, Inc., July 2012.

[COH 73] COHEN-TANNOUDJI C., DIU B., LALOE F., *Mécanique quantique*, Herman, Paris, 1973.

[COL 72] COLELLA R., MENADUE J.F., "Comparison of experimental and beam calculated intensities for glancing incidence high-energy electron diffraction", *Acta Cryst.*, vol. A28, pp. 16–22, 1972.

[COU 89] COURJON D., SARAYEDDINE K., SPAJER M., "Scanning tunneling optical microscopy", *Opt. Comm.*, vol. 71, pp. 23–281, August 2, 1989.

[CUR 80] CURIE J., CURIE P., "Développement, par pression, de l'électricité polaire dans les cristaux hémièdres à faces inclinées", *Comptes rendus de l'Académie des Sciences*, vol. XCI, pp. 294–295, August 2, 1980.

[CUR 81] CURIE P., CURIE J., "Contractions et dilatations produites par des tensions électriques dans les cristaux hémièdres à faces inclinées", *Comptes rendus de l'Académie des Sciences*, vol. XCIII, pp. 1137–1140, December 26, 1981.

[DAH 99] DAHOO P.R., BERRODIER I., RADUCU V. *et al.*, "Splitting of v_2 vibrational mode of CO_2 isotopic species in the unstable trapping site in argon matrix", *Eur. Phys. J. D*, vol. 5, pp. 71–81, 1999.

[DAH 03] Dahoo P.R., Hamon T., Schneider M. *et al.*, "Ellipsometry: Principles, signal processing and applications to metrology", *Proceedings CIMNA*, Lebanon, 2003.

[DAH 04a] Dahoo P.R., Girard A., Tesseir M. *et al.*, "Characterizaton of pulsed laser deposited SmFeO3 morphology: Effect of fluence, substrate temperature and oxygen pressure", *Appl. Phys A, Mat. Sc. Process*, vol. 79, pp. 1399–1403, 2004.

[DAH 04b] Dahoo P.R., Hamon T., Negulescu B. *et al.*, "Evidence by spectroscopic ellipsometry of optical property change in pulsed laser deposited NiO films when heated in air at Neel temperature", *Appl. Phys. A, Mat. Sci. Process*, vol. 79, pp. 1439–1443, 2004.

[DAH 06] Dahoo P.R., Lakhlifi A., Chabbi H. *et al.*, "Matrix effect on triatomic CO_2 molecule: Comparison between krypton and xenon", *J. Mol. Struct.*, vol. 786, pp. 157–167, 2006.

[DAH 11] Dahoo P.R., "Metrological applications of ellipsometry", *Proceedings of 3rd International Metrology Conference 18–23 April 2010 (CAFMET 2010) Cairo, Egypt*, Curran Associates, Inc., September 2011.

[DAH 16] Dahoo P.R., Pougnet P., El Hami A., *Nanometer-scale Defect Detection Using Polarized Light*, ISTE Ltd, London and John Wiley & Sons, New York, 2016.

[DAH 17] Dahoo P.R., Lakhlifi A., *Infrared Spectroscopy of Diatomics for Space Observation 1*, ISTE Ltd, London, and John WIley & Sons, New York, 2017.

[DAH 19] Dahoo P.R., Lakhlifi A., *Infrared Spectroscopy of Diatomics for Space Observation 2*, ISTE Ltd, London, and John WIley & Sons, New York, 2019.

[DAH 21] Dahoo P.R., Lakhlifi A., *Infrared Spectroscopy of Symmetric and Spherical Top Molecules for Space Observation 1*, ISTE, London, and John Wiley & Sons, New York, 2021.

[DIJ 87] Dijkamp D., Venkatesan T., Wu X.D. *et al.*, "Preparation of Y-Ba-Cu oxide superconductor thin films using pulsed laser evaporation from high Tc bulk material", *Appl. Phys. Lett.*, vol. 51, pp. 619–621, 1987.

[DUP 06] Dupas C., Houdy P., Lahmani M., *Nanoscience, Nanotechnologies and Nanophysics*, p. 823, Springer-Verlag Berlin and Heideberg GmbH & Co Eds, 2006.

[FIS 88] Fischer E., Dürig U., Pohl D., "Near-field scanning optical microscopy in reflection", *Appl. Phys. Lett.*, vol. 52, pp. 249–251, 1988.

[FRA 05] FRANTA D., NEGULESCU B., THOMAS L. *et al.*, "Optical properties of NiO thin films prepared by pulsed laser deposition technique", *Appl. Surf. Sci.*, vol. 244, p. 426, 2005.

[GÉR 96] GÉRADIN M., RIXEN D., *Théorie des vibrations : application à la dynamique des structures*, Masson, Paris, 1996.

[HAM 06] HAMON T., BUIL S., POPOVA E. *et al.*, "Investigation of a one dimensional magnetophotonic crystal for the study of ultrathin magnetic layer", *J. Phys. D: Appl. Phys.*, vol. 39, pp. 1–6, 2006.

[HAM 07] HAMON T., Cristaux Magnéto-photonique Unidimensionnels: Etude du Magnétisme de Couches Ultra-minces d'Oxydes, PhD thesis, Paris XI, Orsay, 2007.

[HEC 05] HECHT E. *Optics*, Pearson Education Berlin, 4th edition, 2005.

[HER 45] HERZBERG G., *Molecular Spectra and Molecular Structure II: Infrared and Raman Spectra of Polyatomic Molecules*, volume 2, Van Nostrand D. (ed.), 1945.

[KHE 14] KHETTAB M., *Etude de l'influence du résinage au niveau de L'IML (Insulated Metal Leadframe), dans le packaging de module commutateur de courant mécatronique*, Thesis, UVSQ, Versailles, 2014.

[KIT 80] KITTEL C., KROEMER H., *Thermal Physics*, W.H. Freeman and Company, New York, 1980.

[KNE 20] KNEISSL M., KNORR A., REITZENSTEIN S. *et al.*, *Semiconductor Nanophotonics*, p.556, Springer Nature Switzerland Eds, 2020.

[LAH 04] LAHMANI M., DUPAS C., HOUDY P., *Les nanosciences 1, nanotechnologies et nanophysqiue*, p. 718, Editions Belin, Paris, 2004.

[LAH 06] LAHMANI M., BRECHIGNAC C., HOUDY P., *Les nanosciences 2, nanomatériaux et nanochimie*, p. 26876, Editions Belin, Paris, 2006.

[LAH 07] LAHMANI M., BOISSEAU P., HOUDY P., *Les nanosciences 3, nanobiotechnologies et nanobiologie*, p. 688, Editions Belin, Paris, 2007.

[LAH 10] LAHMANI M., BOISSEAU P., HOUDY P., *Les nanosciences 4, nanotoxicologie et nanoéthique*, p. 608, Editions Belin, Paris, 2010.

[LAN 67] LANDAU L., LIFCHITZ E., *Théorie de l' élasticité*, URSS, Editions MIR, Moscow, 1967.

[LAP 94] LAPLACE P.S., *Oeuvres complètes de laplace*, volume 10, M DCCC XCIV, Editions Académies des Sciences, Gauthiers-Villars et fils imprimeurs de l'Ecole Polytechnique, Paris, 1894.

[LEE 20] LEE Y-C., MOON J-Y, *Introduction to Bionanotechnology*, p. 247, Springer-Verlag, GmbH & Co Editions, Berlin and Heideberg, 2020.

[LEW 83] LEWIS A., ISAACSON M., HAROOTUNIAN A. *et al.*, "Scanning optical spectral microscopy with 500å resolution", *Biophys. J.*, vol. 41, pp. 405a–XXX, 1983.

[LIP 81a] LIPPMANN G., "Principe de la conservation de l'électricité", *Annales de chimie et de physique*, vol. 24, pp. 145–178, 1881.

[LIP 81b] LIPPMANN G., "Principe de la conservation de l'électricité ou second principe de la théorie des phénomènes élecriques", *J. Phys. Theor. Appl.*, vol. 10, pp. 381–394, 1881.

[LOU 16] LOURTIOZ J., LAHMANI M., DUPAS-HAEBERLIN C. *et al.*, *Nanosciences and Nanotechnology: Evolution or Revolution?*, Springer International Publishing Editions, p. 438, 2016.

[LOU 18] LOURTIOZ J., VAUTRIN-UL, C., PALACIN, S. *et al.* (2018). Comprendre les Nanosciences – Session 2, available at: https://www.fun-mooc.fr/courses/course-v1:UPSUD+42003+session02/about, 2018.

[MAR 90] MARSHALL S., SKITEK G., *Electromagnetic Concepts and Applications*, Prentice Hall Inc., Englewood Cliffs, New Jersey, 1990.

[MAX 54] MAXWELL J.C., *A Treatise on Electricity and Magnetism*, 3rd edition, Dover Publications, New York, 1954.

[MEI 15] MEIS C., *Light and Vacuum*, World Scientific Publishing Co., Pte. Ltd, Singapore, 2015.

[MES 64] MESSIAH A., *Mécanique quantique*, vols 1 and 2, Dunod, Paris, 1964.

[MOH 10] MOHSINE A., EL HAMI A., "A robust study of reliability-based optimization methods under eigen-frequency", *Computer Methods in Applied Mechanics and Engineering*, vol. 199, issue 17–20, pp. 17–201006, 2010.

[MOO 65] MOORE G.E., "Cramming more components onto integrated circuit", *Electronics*, vol. 38, no. 8, 1965.

[NOU 07] NOUN W., BERINI B., DUMONT Y. *et al.*, "Correlation between electrical and ellipsometric properties on high-quality epitaxial thin films of the conductive oxide LaNiO3 on STO (001)", *Journal of Applied Physics*, vol. 102, pp. 063709/1–063709/7, 2007.

[NYE 61] NYE J.F., *Propriétés physiques des Cristaux, leur représentation par des tenseurs et des matrices*, translated by BLANC D. and PUJOL T., Dunod, Paris, 1961.

216   Applications and Metrology at Nanometer Scale 1

[POH 84] POHL D.W., DENK W., LANZ M., "Optical stethoscopy: Image recording with resolution $\lambda/20$", *Appl. Phys. Lett.*, vol. 44, pp. 651–653, 1984.

[POH 86] POHL D.W., Optical near field scanning microscope, US Patent 4,604,520, 1986.

[POH 87] POHL D.W., Optical near field scanning microscope, European Patent EP0112401, US Patent US 4604520, May 22, 1987.

[POU 20] POUGNET P., DAHOO P.R., ALVAREZ J.P., "Highly accelerated testing", in POUGNET P., EL HAMI A. (eds), *Embedded Mechatronic Systems 2*, Revised and Updated 2nd Edition, ISTE Press Ltd, London and Elsevier Ltd, Oxford, 2020.

[PRI 62] PRIGOGINE I., *Introduction to Thermodynamics of Irreversible Processes*, John Wiley & Sons Inc, New York, 1962.

[RAT 02] RATNER M., RATNER D., *Nanotechnology: A Gentle Introduction to the Next Big Idea*, Prentice Hall Editions, p. 208, 2003.

[RED 89] REDDICK R.C., WARMACK R.J., FERREL T.L., "New form of scanning optical microscopy", *Phys. Rev. B*, vol. 39, pp. 767–770, 1989.

[RIT 08] RITZ W., "Uber eine neue Methode zur Lösung gewisser Variationsprobleme der mathematischen Physik", *J. Reine Angew. Math.*, issue 135, pp. 1–61, 1908.

[RUS 33] RUSKA E., "Imaging of surfaces which reflect electrons in the electron microscope", *Z. Phys.*, vol. 83, pp. 492–497, 1933.

[SHO 49] SHOCKLEY W., "The theory of p-n junctions in semiconductors and p-n junction transistors", *Bell Syst. Tech. J.*, no. 28, pp. 435–489, 1949.

[SIL 49] SILVER S., *Microwave Antenna Theory and Design*, McGraw-Hill Book Co. New York, USA, 1949.

[SMI 65] SMITH H.M., TURNER A.F., "Vacuum deposited thin films using a ruby laser", *Appl. Opt.*, vol. 4, pp. 147–148, 1965.

[SYN 28] SYNGE E.H., "Suggested method for extending microscopic resolution into the ultramicroscopic region", *Phil. Mag.*, vol. 6, pp. 356–362, 1928.

[SYN 32] SYNGE E.H., "An application of piezoelectricity to microscopy", *Phil. Mag.*, vol. 13, pp. 297–300, 1932.

[TAN 74] TANIGUCHI N., " On the basic concept of 'Nano-Technology'", *Proc. Intl. Conf. Prod. Eng., Part II*, Japan Society of Precision Engineering, Tokyo,1974.

[VOI 10] VOIGT, W., *Lehrbuch der Kristallphysik*. Teubner Verlag, Leipzig, 1910.

[VOI 28] VOIGT, W., *Lehrbuch der Kristallphysik*. Teubner Verlag, Leipzig, 1928.

Index

A, C, D

ablation laser, 17
antenna resistance, 123
atomic force microscope, 6, 25, 30
cathode sputtering, 6–8, 12, 14
diamond, 149
diffraction, 19–21, 26, 27, 29, 46,
    48–50, 92, 93, 205, 207–210
directivity, 115, 123, 133, 135

E, G, I

efficiency, 2, 71, 115, 116, 123, 132
ellipsometry, 10, 14, 17, 32, 35
energy carried by an electromagnetic
    wave, 94
gain, 115, 119, 120, 123, 132, 181
graphite, 149
group velocity, 105, 112
infrared spectroscopy, 35

L, M, N

light, 4, 13–16, 20, 21, 23, 26, 30, 31,
    34, 46, 49, 50, 91–94, 102, 104,
    123, 139, 148, 154, 156, 157, 205
mechatronics, 156
microwave antenna, 94, 114, 134

momentum density of the
    electromagnetic wave, 97
nanometric, 4, 14, 23, 25, 150
nanotechnology, 1, 2
near field, 21, 23, 121, 123, 133, 134

P, R, S

phase velocity, 105, 112
physic-chemical, 6
piezoelectric, 22, 23, 30, 151–156,
    174, 176, 177, 179–181, 190–194,
    197–199, 201
plane wave, 23, 46, 48, 94, 100, 129,
    130, 205, 207
polarization, 93, 94, 99, 115, 116,
    150, 153, 172, 173, 176, 179, 191,
    192, 199
Poynting vector, 94–96, 98, 102, 105,
    112, 114, 123, 130, 134
propagation of uncertainties, 52, 66,
    71
radiated
    field, 129, 139, 144
    power, 116, 118, 120, 123, 131,
        132, 134
radiation diagram, 115, 121, 138,
    139, 143, 145
random, 52–68, 70, 80, 89

reliability, 9, 51, 52, 65, 76, 135
smart materials, 147, 150, 154, 157, 158
SNOM, 23–25
spatial scales, 2
spin coating, 6–10, 33

T, U, W

temporal scales, 3
uncertainty, 4, 51, 65, 70, 79, 80
waveguide, 94, 103–106, 111, 115
wire antenna, 94, 120, 122, 123, 129, 130, 133

Other titles from

in

Mechanical Engineering and Solid Mechanics

2020

SALENÇON Jean
Elastoplastic Modeling

2019

BAYLE Franck
Reliability of Maintained Systems Subjected to Wear Failure Mechanisms: Theory and Applications
(Reliability of Multiphysical Systems Set – Volume 8)

BEN KAHLA Rabeb, BARKAOUI Abdelwahed, MERZOUKI Tarek
Finite Element Method and Medical Imaging Techniques in Bone Biomechanics
(Mathematical and Mechanical Engineering Set – Volume 8)

IONESCU Ioan R., QUEYREAU Sylvain, PICU Catalin R., SALMAN Oguz Umut
Mechanics and Physics of Solids at Micro- and Nano-Scales

LE VAN Anh, BOUZIDI Rabah
Lagrangian Mechanics: An Advanced Analytical Approach

MICHELITSCH Thomas, PÉREZ RIASCOS Alejandro, COLLET Bernard, NOWAKOWSKI Andrzej, NICOLLEAU Franck
Fractional Dynamics on Networks and Lattices

SALENÇON Jean
Viscoelastic Modeling for Structural Analysis

VÉNIZÉLOS Georges, EL HAMI Abdelkhalak
Movement Equations 5: Dynamics of a Set of Solids
(Non-deformable Solid Mechanics Set – Volume 5)

2018

BOREL Michel, VÉNIZÉLOS Georges
Movement Equations 4: Equilibriums and Small Movements
(Non-deformable Solid Mechanics Set – Volume 4)

FROSSARD Etienne
Granular Geomaterials Dissipative Mechanics: Theory and Applications in Civil Engineering

RADI Bouchaib, EL HAMI Abdelkhalak
Advanced Numerical Methods with Matlab® 1: Function Approximation and System Resolution
(Mathematical and Mechanical Engineering SET – Volume 6)
Advanced Numerical Methods with Matlab® 2: Resolution of Nonlinear, Differential and Partial Differential Equations
(Mathematical and Mechanical Engineering SET – Volume 7)

SALENÇON Jean
Virtual Work Approach to Mechanical Modeling

2017

BOREL Michel, VÉNIZÉLOS Georges
Movement Equations 2: Mathematical and Methodological Supplements
(Non-deformable Solid Mechanics Set – Volume 2)
Movement Equations 3: Dynamics and Fundamental Principle
(Non-deformable Solid Mechanics Set – Volume 3)

BOUVET Christophe
Mechanics of Aeronautical Solids, Materials and Structures
Mechanics of Aeronautical Composite Materials

BRANCHERIE Delphine, FEISSEL Pierre, BOUVIER Salima,
IBRAHIMBEGOVIC Adnan
From Microstructure Investigations to Multiscale Modeling:
Bridging the Gap

CHEBEL-MORELLO Brigitte, NICOD Jean-Marc, VARNIER Christophe
From Prognostics and Health Systems Management to Predictive
Maintenance 2: Knowledge, Traceability and Decision
(Reliability of Multiphysical Systems Set – Volume 7)

EL HAMI Abdelkhalak, RADI Bouchaib
Dynamics of Large Structures and Inverse Problems
(Mathematical and Mechanical Engineering Set – Volume 5)
Fluid-Structure Interactions and Uncertainties: Ansys and Fluent Tools
(Reliability of Multiphysical Systems Set – Volume 6)

KHARMANDA Ghias, EL HAMI Abdelkhalak
Biomechanics: Optimization, Uncertainties and Reliability
(Reliability of Multiphysical Systems Set – Volume 5)

LEDOUX Michel, EL HAMI Abdelkhalak
Compressible Flow Propulsion and Digital Approaches in Fluid Mechanics
(Mathematical and Mechanical Engineering Set – Volume 4)
Fluid Mechanics: Analytical Methods
(Mathematical and Mechanical Engineering Set – Volume 3)

MORI Yvon
Mechanical Vibrations: Applications to Equipment

2016

BOREL Michel, VÉNIZÉLOS Georges
Movement Equations 1: Location, Kinematics and Kinetics
(Non-deformable Solid Mechanics Set – Volume 1)

BOYARD Nicolas
Heat Transfer in Polymer Composite Materials

CARDON Alain, ITMI Mhamed
New Autonomous Systems
(Reliability of Multiphysical Systems Set – Volume 1)

DAHOO Pierre Richard, POUGNET Philippe, EL HAMI Abdelkhalak
Nanometer-scale Defect Detection Using Polarized Light
(Reliability of Multiphysical Systems Set – Volume 2)

DE SAXCÉ Géry, VALLÉE Claude
Galilean Mechanics and Thermodynamics of Continua

DORMIEUX Luc, KONDO Djimédo
Micromechanics of Fracture and Damage
(Micromechanics Set – Volume 1)

EL HAMI Abdelkhalak, RADI Bouchaib
Stochastic Dynamics of Structures
(Mathematical and Mechanical Engineering Set – Volume 2)

GOURIVEAU Rafael, MEDJAHER Kamal, ZERHOUNI Noureddine
From Prognostics and Health Systems Management to Predictive
Maintenance 1: Monitoring and Prognostics
(Reliability of Multiphysical Systems Set – Volume 4)

KHARMANDA Ghias, EL HAMI Abdelkhalak
Reliability in Biomechanics
(Reliability of Multiphysical Systems Set –Volume 3)

MOLIMARD Jérôme
Experimental Mechanics of Solids and Structures

RADI Bouchaib, EL HAMI Abdelkhalak
Material Forming Processes: Simulation, Drawing, Hydroforming and
Additive Manufacturing
(Mathematical and Mechanical Engineering Set – Volume 1)

2015

KARLIČIĆ Danilo, MURMU Tony, ADHIKARI Sondipon, MCCARTHY Michael
Non-local Structural Mechanics

SAB Karam, LEBÉE Arthur
Homogenization of Heterogeneous Thin and Thick Plates

2014

ATANACKOVIC M. Teodor, PILIPOVIC Stevan, STANKOVIC Bogoljub, ZORICA Dusan
Fractional Calculus with Applications in Mechanics: Vibrations and Diffusion Processes
Fractional Calculus with Applications in Mechanics: Wave Propagation, Impact and Variational Principles

CIBLAC Thierry, MOREL Jean-Claude
Sustainable Masonry: Stability and Behavior of Structures

ILANKO Sinniah, MONTERRUBIO Luis E., MOCHIDA Yusuke
The Rayleigh–Ritz Method for Structural Analysis

LALANNE Christian
Mechanical Vibration and Shock Analysis – 5-volume series – 3rd edition
Sinusoidal Vibration – Volume 1
Mechanical Shock – Volume 2
Random Vibration – Volume 3
Fatigue Damage – Volume 4
Specification Development – Volume 5

LEMAIRE Maurice
Uncertainty and Mechanics

2013

ADHIKARI Sondipon
Structural Dynamic Analysis with Generalized Damping Models: Analysis

ADHIKARI Sondipon
Structural Dynamic Analysis with Generalized Damping Models: Identification

BAILLY Patrice
Materials and Structures under Shock and Impact

BASTIEN Jérôme, BERNARDIN Frédéric, LAMARQUE Claude-Henri
Non-smooth Deterministic or Stochastic Discrete Dynamical Systems: Applications to Models with Friction or Impact

EL HAMI Abdelkhalak, RADI Bouchaib
Uncertainty and Optimization in Structural Mechanics

KIRILLOV Oleg N., PELINOVSKY Dmitry E.
Nonlinear Physical Systems: Spectral Analysis, Stability and Bifurcations

LUONGO Angelo, ZULLI Daniele
Mathematical Models of Beams and Cables

SALENÇON Jean
Yield Design

2012

DAVIM J. Paulo
Mechanical Engineering Education

DUPEUX Michel, BRACCINI Muriel
Mechanics of Solid Interfaces

ELISHAKOFF Isaac *et al.*
Carbon Nanotubes and Nanosensors: Vibration, Buckling and Ballistic Impact

GRÉDIAC Michel, HILD François
Full-Field Measurements and Identification in Solid Mechanics

GROUS Ammar
Fracture Mechanics – 3-volume series
Analysis of Reliability and Quality Control – Volume 1
Applied Reliability – Volume 2
Applied Quality Control – Volume 3

RECHO Naman
Fracture Mechanics and Crack Growth

2011

KRYSINSKI Tomasz, MALBURET François
Mechanical Instability

SOUSTELLE Michel
An Introduction to Chemical Kinetics

2010

BREITKOPF Piotr, FILOMENO COELHO Rajan
Multidisciplinary Design Optimization in Computational Mechanics

DAVIM J. Paulo
Biotribolgy

PAULTRE Patrick
Dynamics of Structures

SOUSTELLE Michel
Handbook of Heterogenous Kinetics

2009

BERLIOZ Alain, TROMPETTE Philippe
Solid Mechanics using the Finite Element Method

LEMAIRE Maurice
Structural Reliability

2007

GIRARD Alain, ROY Nicolas
Structural Dynamics in Industry

GUINEBRETIÈRE René
X-ray Diffraction by Polycrystalline Materials

KRYSINSKI Tomasz, MALBURET François
Mechanical Vibrations

KUNDU Tribikram
Advanced Ultrasonic Methods for Material and Structure Inspection

SIH George C. *et al.*
Particle and Continuum Aspects of Mesomechanics

Printed in the USA
CPSIA information can be obtained
at www.ICGtesting.com
JSHW011919291223
54426JS00003B/184